# Friction Stir Processing for Enhanced Low Temperature Formability

# Friction Stir Processing for Enhanced Low Temperature Formability

A volume in the *Friction Stir Welding and Processing Book Series*

Christopher B. Smith
Wolf Robotics (formerly of Friction Stir Link)

Rajiv S. Mishra
University of North Texas

AMSTERDAM • BOSTON • HEIDELBERG • LONDON
NEW YORK • OXFORD • PARIS • SAN DIEGO
SAN FRANCISCO • SINGAPORE • SYDNEY • TOKYO

Butterworth-Heinemann is an imprint of Elsevier

Butterworth-Heinemann is an imprint of Elsevier
225 Wyman Street, Waltham, MA 02451, USA
The Boulevard, Langford Lane, Kidlington, Oxford OX5 1 GB, UK

First edition 2014



**Notices**

Knowledge and best practice in this field are constantly changing. As new research and experience broaden our understanding, changes in research methods, professional practices, or medical treatment may become necessary.

Practitioners and researchers must always rely on their own experience and knowledge in evaluating and using any information, methods, compounds, or experiments described herein. In using such information or methods they should be mindful of their own safety and the safety of others, including parties for whom they have a professional responsibility.

To the fullest extent of the law, neither the Publisher nor the authors, contributors, or editors, assume any liability for any injury and/or damage to persons or property as a matter of products liability, negligence or otherwise, or from any use or operation of any methods, products, instructions, or ideas contained in the material herein.

**British Library Cataloguing-in-Publication Data**
A catalogue record for this book is available from the British Library

**Library of Congress Cataloging-in-Publication Data**
A catalog record for this book is available from the Library of Congress

ISBN: 978-0-12-420113-2

For information on all Butterworth-Heinemann visit our website at store.elsevier.com

This book has been manufactured using Print On Demand technology. Each copy is produced to order and is limited to black ink. The online version of this book will show color figures where appropriate.

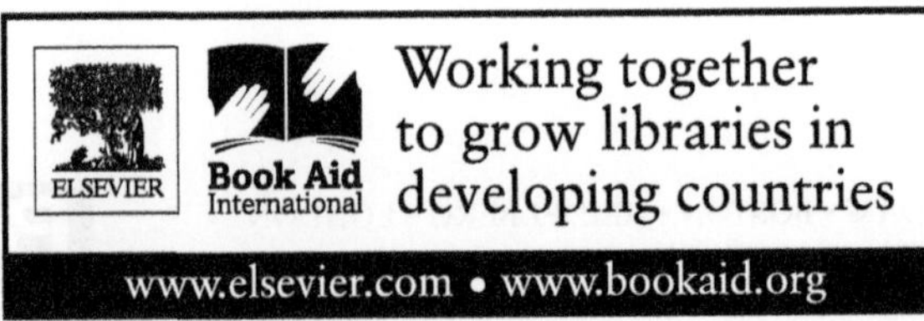

# CONTENTS

**Acknowledgments** ........ **vii**
**Preface** ........ **ix**

**Chapter 1 Concept of Friction Stir Processing for Enhanced Formability** ........ **1**
1.1 Background ........ 1

**Chapter 2 Fundamentals of Formability** ........ **7**
2.1 Introduction ........ 7
2.2 Tensile Test and Formability ........ 7

**Chapter 3 High Structural Efficiency Design Potentials with Enhanced Formability** ........ **11**
3.1 Background ........ 11
3.2 Fabrication Processes ........ 11
3.3 Summary ........ 18

**Chapter 4 Case Study of Aluminum 5083-H116 Alloy** ........ **19**
4.1 Case Study Initiation ........ 19
4.2 Initial Feasibility Results ........ 20
4.3 Case Study Description ........ 24
4.4 Initial Comparative Process Qualification ........ 28
4.5 GMAW Qualification and Destructive Testing ........ 31
4.6 FSP Qualification and Destructive Testing ........ 34
4.7 FSW Initial Qualification and Destructive Testing ........ 34
4.8 Macro Cross Section Comparison ........ 35
4.9 Microhardness Test Results and Comparison ........ 39
4.10 Macro Transverse Tensile Test Results ........ 44
4.11 Young's and Shear Modulus Test Results ........ 49
4.12 Mini Transverse Tensile Test Results ........ 51
4.13 Mini Longitudinal Tensile Test Results ........ 75
4.14 Distortion Measurements ........ 86
4.15 Corrosion Testing ........ 94
4.16 Fatigue Testing ........ 115
4.17 Corrosion Fatigue Testing ........ 122

**Chapter 5 Examples of Enhanced Formability of High-Strength Aluminum Alloys .......... 125**

5.1 Background .......... 125

5.2 Examples of Enhanced Formability of High-Strength Aluminum Alloys .......... 125

5.3 Summary .......... 131

**References .......... 133**

# ACKNOWLEDGMENTS

This project was funded by the U.S. Navy under contract N00014-08-C-0089. The prime contractors for this work were Friction Stir Link, Inc. (FSL) and the Missouri University of Science and Technology (MS&T). The U.S. Navy's Naval Surface Warfare Center – Carderock Division – provided guidance and technical oversight. All of the welding and processing was performed by Friction Stir Link, Inc. with material property testing primarily being completed by MS&T. Other material property testing was performed by FSL, the South Dakota School of Mines and Technology (SDSMT), and a limited amount at material property testing suppliers.

The authors would like to sincerely thank all who have provided contributions that made this short book possible. First of all, thank you to the U.S. Navy for providing the monetary resources that have allowed a very comprehensive, expansive, and likely the largest data set to date, comparing friction stir processing, friction stir welding, gas metal arc welding, and base material. However, this would not have been possible without the foresight of Bruce Halverson and Scott Craw at Marinette Marine, as well as Maria Posada and Johnnie DeLoach at the U.S. Navy, with their ability to envision the benefits of the use of friction stir processing and friction stir welding. With their support and technical guidance, this project and the data presented were created by a large team that includes colleagues at FSL, MS&T, and SDSMT. Major contributions to this work included, but were not limited to Adam O'brien, Lee Cerveny, and Jerry Opichka at FSL, as well as Arun Mohan, Jianqing Su, Gaurav Argade, and Kumar Kandasamy at MS&T. Furthermore, significant contributions and guidance were provided by Murray Mahoney (retired and formerly of Rockwell Scientific) as well as Professor Mike Miles from Brigham Young University. Next, significant contributions were also provided by the team SDSMT for their structural testing contribution for which the authors would specifically like to thank Damon Fick and Christian Widener for their contributions. Providing additional guidance, the authors would also like to acknowledge the contribution of Cathy Wong, Kirsten Green, and Nat Nappi of the U.S. Navy for their technical guidance on the testing methods and approaches that would provide the most valuable comparison data with the traditional approach. Finally, the authors would like to specifically recognize John F. Hinrichs of Friction Stir Link who was a great champion of this work and always encouraged advancement of welding and its related technologies and whose contributions to the welding community will be missed. This was truly a team effort. Thanks again to all. . .your efforts have all been very much appreciated!

# PREFACE

This is the second volume in the recently launched short book series on friction stir welding and processing. As highlighted in the preface of the first book, the intention of this book series is to serve engineers and researchers engaged in advanced and innovative manufacturing techniques. Friction stir processing was started as a generic microstructural modification technique almost 15 years back. In that period, friction stir processing-related research has shown wide promise as a versatile microstructural modification technique and solid-state manufacturing technology. Yet, broader implementations have been sorely missing. Disruptive technologies face greater barrier to implementation as designers do not have these in their traditional design tool box! Part of the inhibition is due to lack of maturity and availability of large data set.

This book has primarily a technological flavor. It contains significant volume of data generated as a part of technology approval document for the US Navy. The intention behind this book is to share an example that can boost the confidence with engineers and designers as they consider friction stir processing as a viable technology for advanced manufacturing. This short book series on friction stir welding and processing will include books that advance both the science and technology.

*Rajiv S. Mishra*
*University of North Texas*
*March 8, 2014*

# CHAPTER 1

# Concept of Friction Stir Processing for Enhanced Formability

## 1.1 BACKGROUND

Since its invention in 1991 and then first production implementation in 1995, friction stir welding (FSW) has experienced a continual increase in use throughout the world (Thomas et al., 1991; Dawes and Thomas, 1995). Its increase in implementation has occurred because of FSW's benefits over traditional fusion welding processes, such as improved static strength, improved fatigue properties, less sensitivity to disturbances (e.g., contamination), and significantly less distortion. The benefits of FSW are arguably the most prominent when comparing with fusion welding of aluminum. As such, most of the implementation of FSW has occurred in the aluminum fabrication industry, especially for applications that have been specifically designed for FSW.

The FSW process is fairly simple in concept and is illustrated in Figure 1.1 (Mishra and Ma, 2005). The FSW process first involves a machine initiating rotation of a friction stir tool. The friction stir tool has a probe and a shoulder, both with specially designed features. The FSW machine then plunges the rotating friction stir tool into the workpiece, creating heat locally via friction and plastic deformation of the material, softening the material to be welded. Once the probe is completely plunged into the workpiece and the shoulder contacts the face surface, the FSW machine initiates the traverse of the friction stir tool along the weld path or joint. While the FSW machine traverses the tool along the path, the rotation of the tool is maintained, with geometric features on the shoulder and probe displacing and mixing (i.e., stirring) material along the weld joint. Then, when the friction stir tool reaches the end of the path, the friction stir tool is retracted from the joint and finally rotation of the friction stir tool is ceased.

During the initial years of research and implementation, it was observed that FSW would locally modify the material properties in and around the weld area. As FSW is an autogenous process

Friction Stir Processing for Enhanced Low Temperature Formability. DOI: http://dx.doi.org/10.1016/B978-0-12-420113-2.00001-5

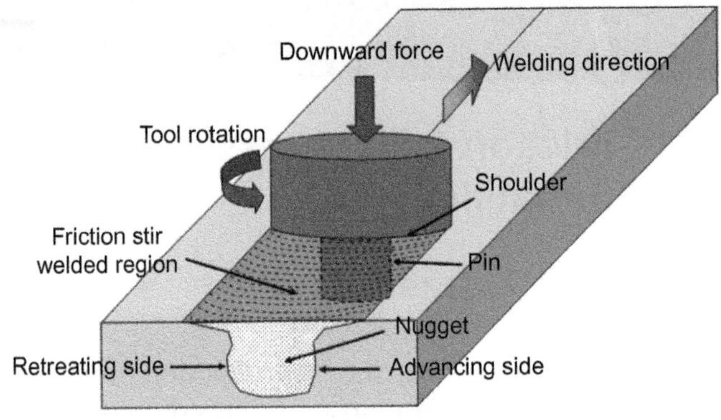


*Figure 1.1 Schematic illustration of friction stir welding.*

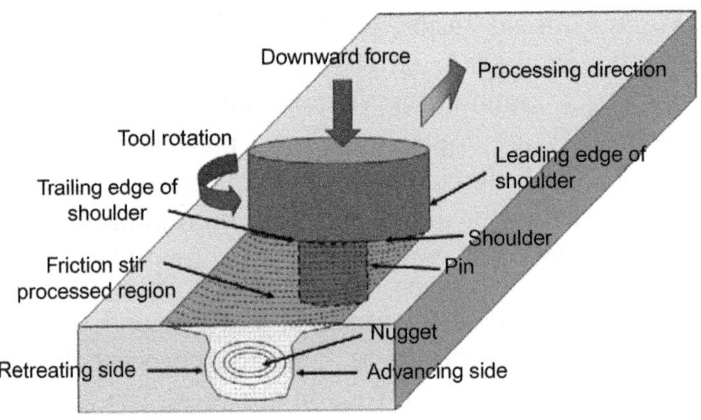


*Figure 1.2 Schematic illustration of FSP.*

(no filler material), the area of the weld has chemistry identical to the base material. With these two considerations, a variant of the FSW process was developed and is referred to as friction stir processing (FSP) (Mishra et al., 1999; Mishra and Mahoney, 2001). The FSP basically involves the same concept as FSW, but it is generally performed on base material, without a weld joint and is shown in Figure 1.2. FSP has been demonstrated to be capable of locally modifying various material properties, including but not limited to ductility/elongation, fatigue properties, static mechanical properties, corrosion properties, hardness, etc. (Mishra and Mahoney, 2007). The ability to modify material properties is material dependent. As such, FSP can be used to locally improve material properties to enhance product capability or to enable different fabrication methods. While there are numerous potential applications, a few such applications include:

1. Enhancing the ductility of various aluminum alloys, enabling forming of components versus the traditional approach of fabrication (welding) of a number of detail components.
2. Improving hardness of various steel alloys. FSP can be used to displace the traditional hardness improvement processes such as heat treating, which are energy intensive. In addition, heat treatment processes tend to be global in nature, whereas FSP can be used to customize surface properties based on location.
3. Elimination of porosity and improvement of mechanical properties in cast materials.

Specific to enhanced formability, potential applications exist in three broad areas:

1. FSP of sheet or plate to enhance or enable superplastic forming (high temperature forming). FSP has been demonstrated to be able to impart superplastic behavior (elongations >200%) in several aluminum alloys. Potential applications in this area would tend to allow replacement of a multicomponent assembly with a formed sheet. Sheet thickness in these applications would tend to be below 6 mm although FSP conceptually opens possibilities of higher thicknesses. Research has indicated that the most viable applications would have volumes in the several hundred to several thousand per year, though recent reductions in cost through automation of superplastic forming has allowed superplastic forming and its variants to be commercially viable in higher production volume application. In addition, components with higher part counts using traditional fabrication approaches would tend to have better commercial viability. An example of an application is door structures for various industries. One such example is a ship door shown in Figure 1.3.
2. FSP of thin plate, followed by room temperature bending. Potential applications in this area could include roll forming or press brake forming of angle, C-channels, and other shapes. These shapes would otherwise be extruded or fabricated (fusion welded) from multiple sections of thin plate. With marine grade Al alloys (typically 5XXX alloys) or with low volumes, extrusions tend to be

*Figure 1.3 An example of a ship door.*

costly making this alternative approach very beneficial in marine fabrication and other applications where 5XXX alloys are used. Plate thicknesses in these applications would tend to be between 6 (1/4″) and 13 mm (1/2″). These applications could additionally be characterized as requiring only a single FSP pass. Applications for such a process would be structural components used in multiple industries ranging from marine, truck trailer, rail car, etc. An example of a section of a structural component (an angle) FSP and then formed is shown in Figure 1.4.

3. FSP of thick plate, followed by room temperature bending. In these applications, fusion welding of multiple sections of thick plate could be replaced by FSP and then a forming operation. Potential applications in this area could include major marine structural members or aluminum armored vehicles. The alternative is especially attractive in armor applications, as ballistic properties would not be affected negatively, unlike the traditional fusion welding approach. Furthermore, significant cost reductions could be realized. Plate thicknesses in these

*Figure 1.4 An example of a section of a structural component (an angle) FSP and then formed.*

*Figure 1.5 An example of an FSP (multiple passes) and formed thick plate.*

applications would tend to be in excess of 13 mm (1/2″) and would require multiple pass FSP. An example of an FSP (multiple passes) and formed thick plate is shown in Figure 1.5.

With all the twenty plus years of research that has been performed on FSW and FSP, almost all of the research has focused on comparing limited and specific material properties of FSW or FSP of a specific alloy versus base material or other traditional processes, such as gas metal arc welding

(GMAW). GMAW is often a basis of comparison, given its wide use in the joining of aluminum and steel structures. With this, there has been no comprehensive case study, review, or investigation comparing material properties of FSW and FSP versus traditional processes and/or the base material. This short book is a summary of a case study of a large project comparing and contrasting the FSP with GMAW and the base material for a large and comprehensive set of material properties.

CHAPTER 2

# Fundamentals of Formability

## 2.1 INTRODUCTION

Sheet metal forming is quite common for making shaped components, from soda cans to automotive car bodies. It is customary to refer to a material below the thickness of 6.35 mm as a sheet and thicker materials as plate. Because this book is limited to bend forming, which is the simplest of sheet metal forming operations, the basic discussion is limited to bending. It is easy to visualize the extension of friction stir processing for enhanced formability to three-dimensional forming involving various stress states. A key aspect that is going to be emphasized is the plate forming. As will become clear, higher the thickness of material, lower the formability.

Figure 2.1 shows three basic types of sheet metal deformation associated with sheet metal forming and basic terminology associated with sheet metal bending. The ratio of bend radius to sheet thickness is critical and defines how aggressive the local strain gradient is. The simplest mechanical test is the tensile testing of sheet materials. In the next section, the basics of formability are discussed in the context of bending of sheets and plates.

## 2.2 TENSILE TEST AND FORMABILITY

The tensile test is the simplest form of mechanical testing and can yield important parameters for discussion of formability. The basic equations governing deformation are given as

$$\sigma = k\varepsilon^n \tag{2.1a}$$

$$\sigma = k(\varepsilon - \varepsilon_o)^n \tag{2.1b}$$

$$\sigma = k(\varepsilon - \varepsilon_o)^n \dot{\varepsilon}^m \tag{2.1c}$$

where $\sigma$ is the flow stress, $\varepsilon$ is the flow strain, $k$ is a constant, $n$ is the strain hardening coefficient, $\varepsilon_o$ is a reference strain, $\dot{\varepsilon}$ is the strain rate, and $m$ is the strain rate sensitivity. Equation (2.1c) has a more generic form that takes care of strain hardening as well as strain rate

Friction Stir Processing for Enhanced Low Temperature Formability. DOI: http://dx.doi.org/10.1016/B978-0-12-420113-2.00002-7

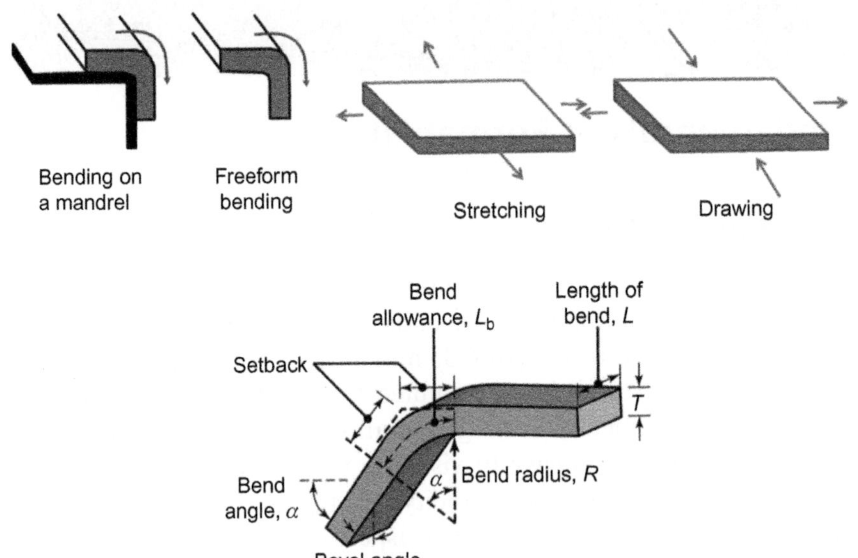


*Figure 2.1 Schematic of three basic sheet metal forming with type of deformation. The terminology of bending (Kalpakjian and Schmid, 2006) is included. The ratio of bend radius and sheet thickness is critical.*

dependence of flow stress. The strain hardening coefficient and strain rate sensitivity are important terms for the stability of flow and onset of instability or failure mechanisms.

Datsko and Yang (1960) established a very simple relationship between tensile properties and bending as

$$\frac{R}{t} = \frac{50}{A_r} - 1 \tag{2.2}$$

where $R$ is the radius of bend (also referred as bend radius), $t$ is the thickness of sheet/plate, and $A_r$ is the reduction of area during a tensile test. Figure 2.2 shows the variation of $R/t$ against $A_r$. This highlights the importance of ductility and reduction of area of the sheet or plate. Friction stir processing increases the ductility and reduction of area of a material. This directly increases the bend formability.

More recently, forming limit diagrams are used to depict the safe zone (Ghosh and Hecker, 1974). Figure 2.3 shows such a forming limit diagram taken from the textbook of Kalpakjian and Schmid (2006). The plane strain condition applies during bending of sheet and plates. It can be noted that this has the minimum safe-limit strain.

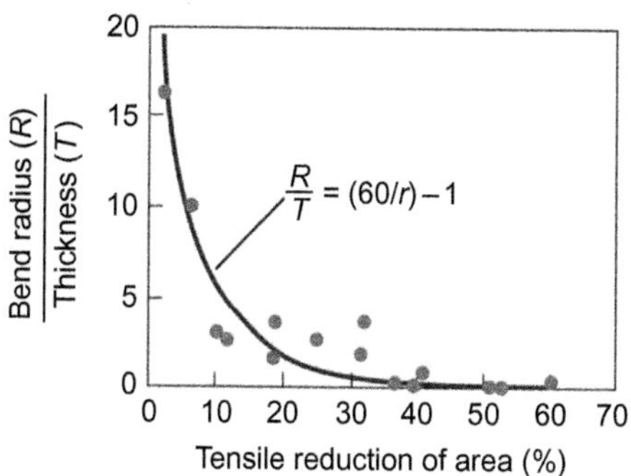


*Figure 2.2 Variation of* R/t *with reduction of area.* After Datsko and Yang (1960).

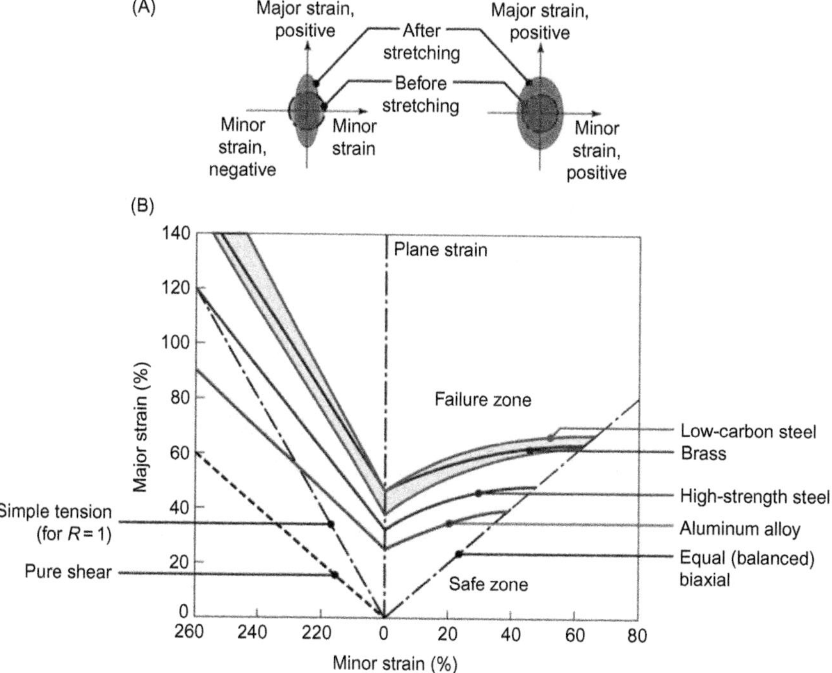


*Figure 2.3 (A) Forming limit diagrams used to depict the safe zone. (B) Bending of sheet or plate involves plane strain.* Figure taken from Kalpakjian and Schmid (2006).

# CHAPTER 3

# High Structural Efficiency Design Potentials with Enhanced Formability

## 3.1 BACKGROUND

Angles, C-channels, zees, rectangular tubes, among others are all shapes that are used as detail components in structural fabrications. Furthermore, any of the listed shapes can employ friction stir welding (FSW), friction stir processing (FSP), and/or gas metal arc welding (GMAW) in the fabrication of such shapes (Figure 3.1).

These structural angle components can be formed to a right angle with a single bend, but with multiple FSP and bending operations, other shapes can be fabricated, such as C-channels or zees. FSP/bending allows long-length sections (limited only by the length of the plate) to be fabricated. This new approach can lead to significant cost reductions, improved material properties, and improved component performance as will be discussed in the remainder of this book.

## 3.2 FABRICATION PROCESSES

### 3.2.1 Traditional Approach

With the traditional approach, via use of GMAW, there are multiple approaches to fabrication of structural angles. The two most viable approaches are discussed in this section. The simplest approach first involves creation of two strips, with width equal to the desired leg lengths of the angle, as shown in Figure 3.2. The angles are then prepared for GMAW. Typically, a bevel would be generated on one of the sections, especially if the thickness of the legs of the angle is above a certain threshold. The bevel would be placed on the inside corner of the joint. The joint would then be welded. A single weld to multipass welds can be required depending on the thickness of the angle. Once the joint is completed, the back side of the joint (outside of the angle) would be back gouged for preparation for a final weld pass. After this, the final weld pass is completed to ensure a full penetration joint. Lastly, the ends may be removed to cut the part to its final length, so

Friction Stir Processing for Enhanced Low Temperature Formability. DOI: http://dx.doi.org/10.1016/B978-0-12-420113-2.00003-9

*Figure 3.1 Example of angles used in structural fabrication.*

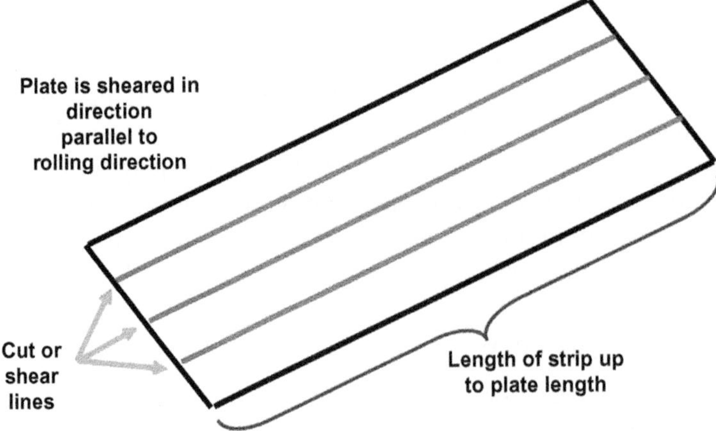


*Figure 3.2 In the traditional welding approach, the plate is cut or sheared in strips in the lengthwise direction.*

as to remove any weld of suspect quality at the start and stop and to eliminate tolerance issues associated with staggered parts. This concept is shown in Figure 3.3. Given the amount of welding per length of part, a relatively high level of distortion can be expected.

Due to the high level of distortion that can be generated with the approach described earlier, another approach is often considered. This first involves shearing or cutting plate into shorter strips, as shown in Figure 3.4. These short strips are then formed/press braked with a large radius, as displayed in Figure 3.5. The strips are limited to 4–8 ft in width, which is generally equivalent to the width of the plate that is supplied to the shearing or cutting operation. The plate is cut in this

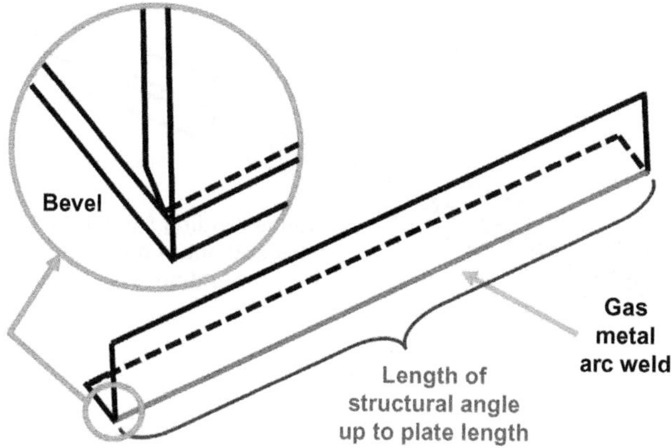


*Figure 3.3 The pieces are gas metal arc welded in the two-strip approach.*

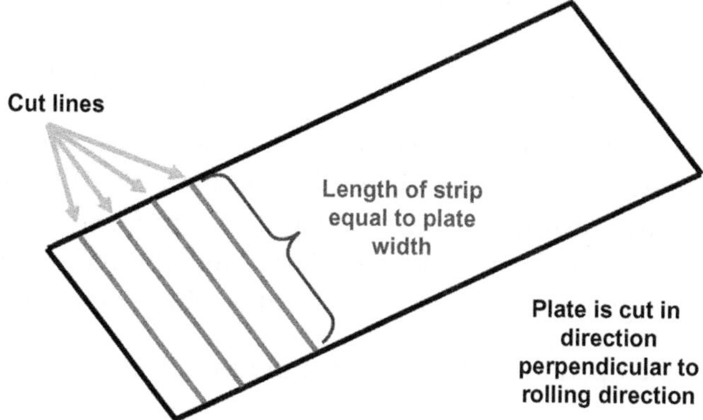


*Figure 3.4 Illustration of cut or shear plate in strips widthwise.*

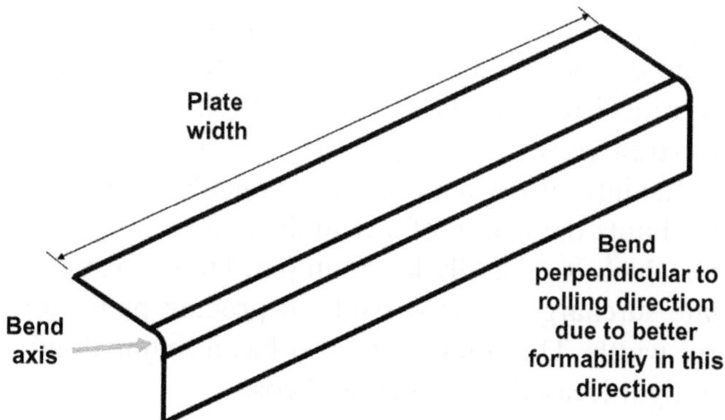


*Figure 3.5 Illustration of formed/press braked splice with a large radius.*

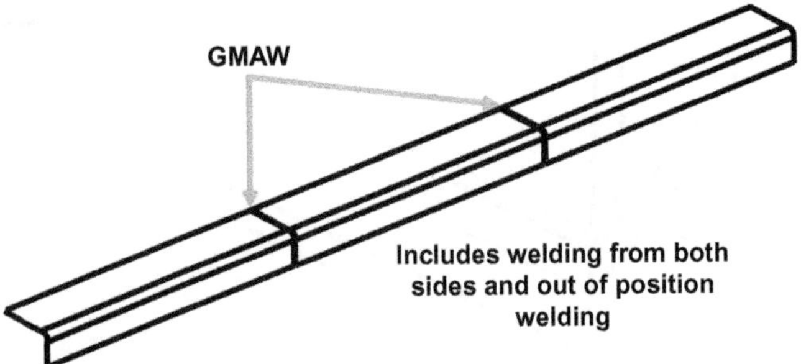


*Figure 3.6 Illustration of welding of short formed splices.*

direction because the formability of the aluminum is significantly better when the bend axis is perpendicular to the rolling direction of the material. Forming with the bend axis parallel with rolling direction is even more severely limited (Kalpakjian and Schmid, 2006). It is also noted that material costs (e.g., dollars/pound) increase with plate above a 4 ft width, creating a cost penalty for strips longer than 4 ft.

These formed short sections are then spliced to create a longer angle, as shown in Figure 3.6. The splice occurs every 4–8 ft (equivalent to the width of rolled plate). As with the approach via fabrication with two long strips, there is first GMAW preparation activity, welding, then back-gouging, followed by a final welding operation.

The specific operations are described in more detail as such:

1. *Operation 1*: The first operation involves cutting of flat plate or sheet into strips to a predetermined width as shown in Figure 3.2. This is typically performed via shearing in a press brake, water jet process, plasma process, or a laser cutting process. The process used will depend on the thickness of the plate and the desired edge quality. It is critical to note that the plate or sheet is cut in the width direction (transverse to rolling direction) of the plate, due to the superior ductility of the material perpendicular to the rolling direction. This limits the length of any individual section to the width of the plate, which is generally less than 6 ft. However, with significant wide plate surcharges, it is possible to procure plates up to 8 ft or greater in width. This wide plate also has a major disadvantage of significantly increased lead times and cost.
2. *Operation 2*: Forming of the structural angle to a right angle with a relatively large radius, to avoid cracking of the aluminum, as shown

in Figure 3.5. This is typically performed in a press brake of length capacity at least equal to the width of the plate (length of the structural angle segment). The forming is typically performed at radii approximately four times the thickness (4T) of structural angle. For example, structural angles fabricated from material 8 mm in thickness are formed at a 31 mm radius.

3. *Operation 3*: The third operation involves preparation of the splice joints for GMAW. Typically, a bevel is generated to create a groove joint, especially if the plate is above a certain thickness threshold. The joint is then cleaned prior to the welding operation.
4. *Operation 4*: The next operation is to splice the individual sections via GMAW to create a structural angle longer than the width of the plate, as shown in Figure 3.6. The splicing first involves GMAW from the side in which the bevel was created. The part will then be flipped, back gouged, cleaned, and then the gas metal arc weld is made from the reverse side. It is also noted that there is often a repositioning of the part in between welding of each of the legs from each side, so as to keep the welding in the 1G position (weld in horizontal plane). Furthermore, run-on and run-off tabs are often used so as to improve the weld quality at the starts and stops.
5. *Operation 5*: The last operation involves grinding of the welds to remove any convexity. As with the welding, the parts need to be repositioned to gain optimal axis for manual grinding.

As with the two-strip approach, the number of weld passes and joint preparation are a function of the thickness of the angle that needs to be fabricated. This secondary approach significantly reduces the amount of welding, reducing the overall amount of distortion. However, the welding is more difficult due to the requirement for out of position welding orientations due to welding around the large bend radius. Thus, this approach is somewhat more susceptible to weld quality issues.

### 3.2.2 FSP and FSW Approach

The proposed approach to fabricate these large structural angles is also a multiple step process, consisting of the following operations:

1. *Operation 1*: The first operation involves cutting of flat plate or sheet into strips along the length of the plate to a predetermined width as shown in Figure 3.2. This is typically performed via shearing in a press brake, water jet process, plasma process, or a laser

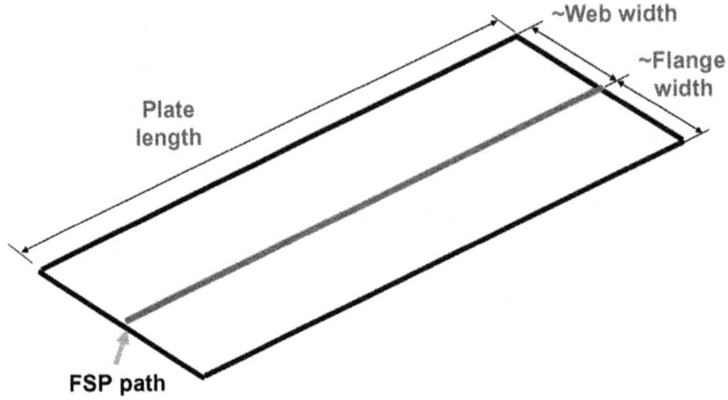


*Figure 3.7 FSP of the strip in the flat condition along the full length of the future bend line.*

cutting process. The process used will depend on the thickness of the plate and the desired edge quality. It is critical to note that the plate or sheet is cut in the length direction (parallel to rolling direction) of the plate, which is opposite to that of the traditional approach. This limits the length of any individual section to the length of the plate, which is typically a minimum of 20 ft. Plates of up to 40 ft in length can be procured, but like the wide plate, there are increased lead times. However, the surcharges are generally less than for wide plate.

2. *Operation 2*: The second operation involves FSP of the strip in the flat condition along the full length of the future bend line, as shown in Figure 3.7. FSP uses a single pass, partial penetration tool (<40% of the plate thickness) at a predetermined travel speed and tool rotation rate. In the friction stir processed zone, the microstructure will be an equiaxed, fully recrystallized fine grain microstructure. Further, in the volume directly influenced by the FSP tool, the microstructure will be an F temper, that is, as-fabricated with no special control of thermal conditions or strain hardening. Adjacent to the FSP zone, there will be a conventional heat affected zone (HAZ) and unique to FSP, a very small thermomechanical affected zone. However, the HAZ volume will be less than that experienced with fusion welding due to the reduced heat input, shorter thermal cycle, and the shallow tool penetration. As part of this operation, the friction stir processed surface will be ground after FSP. There is little to no flash, but the FSP tool does leave a shallow (<0.001 in. deep) swirl pattern on the surface. This pattern is generally not deleterious, but is typically removed by grinding prior to bending.

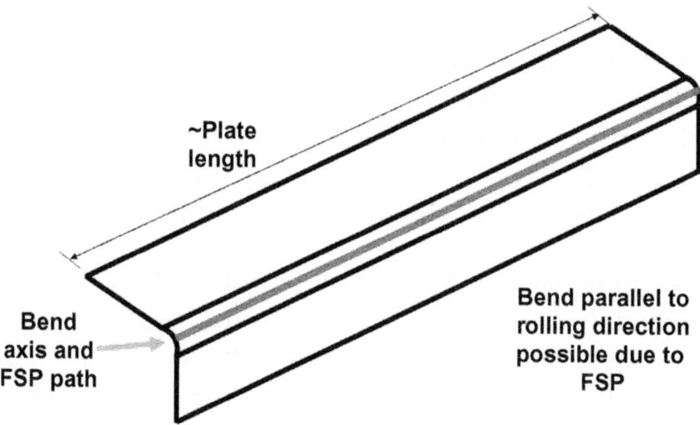


*Figure 3.8 Forming of the friction stir processed strip to a right angle with a relatively small radius.*

3. *Operation 3*: The third operation involves saw cutting the ends of the strip to remove the FSP starts and stops.
4. *Operation 4*: The last operation involves forming of the friction stir processed strip to a right angle with a relatively small radius, as shown in Figure 3.8. Bending for this application is accomplished in a press brake, though roll forming could be used. Selecting between press brake and roll forming requires considerations of cost, technical issues, and practical concerns. For example, roll forming accommodates very long lengths, whereas press brake forming is limited by the width of the press brake. Also, if important, roll forming, because it is progressive and is in longer contact with the dies, has less springback and thus more repeatable dimensional accuracy in the final part and less initial scrape. Further, the tooling costs for roll forming are higher. With that, unless the production quantity is large, press break forming is more cost effective. Each approach can accommodate different thicknesses.

This simple multistep fabrication approach creates a homogeneous microstructure and thus homogeneous mechanical properties along the structural angle length. However, in a cross-section of a structural angle fabricated via FSP, there will be different properties for the different thermomechanical histories, that is, parent metal for most of the cross-section, a small HAZ, and the small strain hardened FSP zone. A smaller bend radius has the benefit of increasing the section modulus of the component, increasing the strength and reducing the stress level in the assembled panel.

One of the limitations of the FSP approach described earlier versus the traditional approach is that the length of the structural angle is limited to the length of the plate. However, this can be resolved via splicing, as well. However, splicing with GMAW can negate the benefits of the FSP approach. To avoid this concern, the strips can be spliced prior to forming with FSW, eliminating the limitation on length.

From a commercial or production operations viewpoint, the FSP along the bend line basically substitutes the splicing operation via GMAW of the traditional approach. Otherwise, the fabrication approaches are fairly similar. The splicing operation of the traditional approach is very time consuming, whereas the FSP requires a fraction of the time. This leads to significant potential cost reductions for the proposed approach versus the traditional approach.

Furthermore, the traditional approach creates a more difficult to manage level of distortion. This results in fit-up issues when joining the structural angle to the adjoining structures. This causes nonvalue-added costs that are associated with the traditional approach that can be eliminated or significantly reduced with use of the FSP approach, further reducing fabrication costs. To summarize, the new approach with FSP eliminates the need for the GMAW splices, produces a long-length structure with homogeneous properties along the length, reduces distortion, and eliminates wide plate surcharges.

## 3.3 SUMMARY

The case study used in this book is a summary of a large project investigating a multitude of material properties (19 in total) of FSW, FSP versus GMAW, and the base material for a single aluminum alloy (5083-H111). 5083-H111 was selected as the material of choice, since it is used in applications where FSW, FSP, and GMAW can all be used to fabricate structural components. In this study, a fabricated angle was chosen as the design of a component for the comparisons, given the ability to use all processes on this design and the wide use of structural shapes in a variety of industries. This book compares the local and global material properties of aluminum angles fabricated with the various processes mentioned. The geometric design of the aluminum angle was kept as common as possible for the various processes.

# CHAPTER 4

# Case Study of Aluminum 5083-H116 Alloy

## 4.1 CASE STUDY INITIATION

Given its simplicity and being structurally shaped that can employ friction stir welding, friction stir processing, and gas metal arc welding in its fabrication process, the angle represents a good candidate to compare and contrast the various processes. An example of such an angle in a structural application is shown in Figure 4.1. In structural applications, 5083-H116 is commonly selected for several reasons. The first is that this alloy, and its close variant, 5086, have excellent corrosion resistance. Second, this alloy has superior corrosion resistance to other structural alloys. Lastly, equivalent-shaped extrusions of the geometry needed for most applications are generally not possible in this alloy. With these characteristics, this alloy is commonly used in the marine, rail car, and heavy truck industries.

These structural angle components can be formed to a right angle with a single bend but with multiple FSP and bending operations other shapes can be fabricated, such as c-channels or zees. FSP/bending allows long length sections (limited only by the length of the plate) to be fabricated. This new approach can lead to significant cost reductions, improved material properties, and improved component performance as will be discussed in the remainder of this book.

With the described potential benefits and wide application, in 2006, a Phase I STTR (Small Business Technology Transfer Program) project entitled "Friction Stir Processing to Improve Formability of Aluminum," was awarded to Friction Stir Link, Inc. with the sponsoring agency or contractor being the US Navy. As a part of this project several applications were identified, where FSP could potentially be used to locally enhance the ductility of aluminum, providing both technical improvements and cost reductions. One such application was the use of FSP to enable room temperature bending of aluminum for marine structural components including stiffeners, frames, and girders. After review of potential marine applications, multiple applications for structural right angles of the design noted in the previous sections were

**Friction Stir Processing for Enhanced Low Temperature Formability. DOI: http://dx.doi.org/10.1016/B978-0-12-420113-2.00004-0**

*Figure 4.1 An example of an angle in a structural application and a view of the overall structure.*

identified. These structural right angles have a range of thickness but are typically greater than 3 mm and less than 16 mm for most marine applications.

As a part of Phase II of the noted project, technical feasibility trials were performed to understand the benefits or feasibility of using FSP to locally enhance the room temperature ductility of aluminum. As noted, multiple potential applications were identified. However, technical feasibility trials were only performed for one structural angle application.

## 4.2 INITIAL FEASIBILITY RESULTS

For the feasibility trials, 8-mm-thick (5/16″) aluminum alloy AA 5083-H116 was selected to demonstrate technical feasibility for this application. This particular material was chosen due to a significant use of this alloy throughout marine industry for aluminum-based ships. The 8 mm thickness was chosen, since it was at the mid-range of the thickness of potential applications in the marine industry and it represented a thickness where significant work content could be envisioned. Furthermore, it was estimated that a single pass friction processing technique could be used to improve a sufficiently large area or width to allow for forming. If the thickness of the material was sufficiently high, then more than one friction stir processing pass would be required. This process development first involved design and fabrication of several different friction stir tools. Processing trials were performed at various FSP parameters, including multiple rotation speeds and travel speeds. The friction stir processed sections were tensile

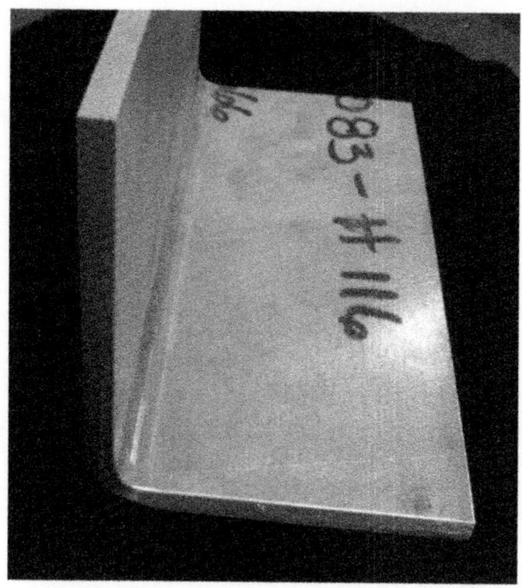

*Figure 4.2 An example of a formed sample with small bend radius.*

tested and wraparound bend tested to determine the quality of the FSP sections, the relative capability of FSP to enhance the ductility, and to determine to what radii the aluminum could be formed. As a result of these tests, it was apparent that the FSP could potentially be used to allow 8-mm AA 5083-H116 to be formed to inside bend radii to as small as 6 mm, versus greater than 25 mm without FSP. From these initial process development trials, nominal friction stir processing parameters were selected.

In this study, once nominal FSP parameters were determined with an associated nominal bend radii, additional small samples were created. Some of these samples were left as is and some were formed in a press brake to a 6 mm inside radius. One such formed sample is shown in Figure 4.2.

Metallurgical analysis and mini tensile testing was performed at the locations shown in Figure 4.3. The red dots indicate locations where mini testing was performed for all samples and the green dots indicate locations where a smaller number of samples were tested. Results of the metallurgical analysis (cross sections etched with Keller's reagent) at each of the red-dot locations are shown in Figure 4.4. As shown, the material within the friction stir processed zone has a much finer grain structure.

Mini tensile testing was performed at the locations indicated in Figure 4.3, with the testing direction being transverse to the processing

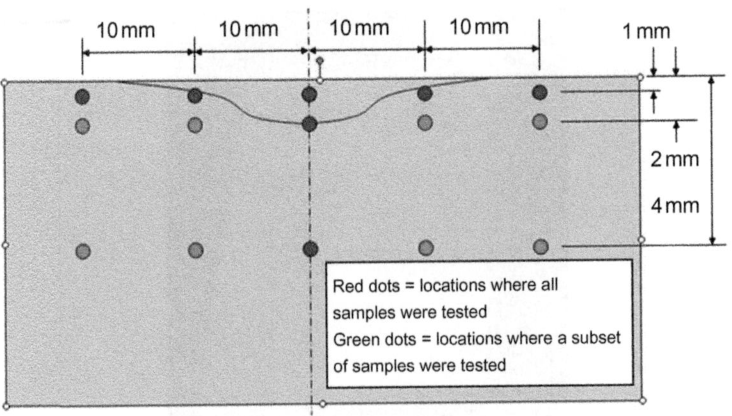


*Figure 4.3 Locations of metallurgical analysis and mini tensile testing.*

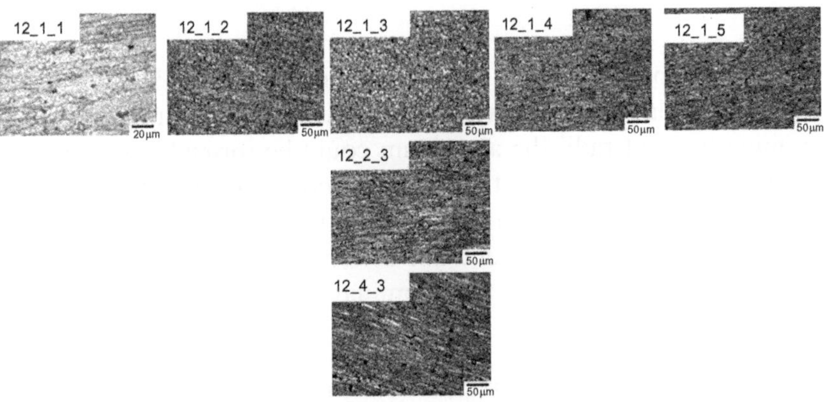


*Figure 4.4 Microstructure locations from the locations marked in Figure 4.3.*

direction. As with the cross sectioning, this testing was performed on all samples at the locations indicated by the red dots and on a subset of samples at the locations indicated by the green dots. This was performed both before and after forming. The tensile test coupons had a gage length of 2 mm, gage thickness of 0.5 mm, and a width 1.0 mm. Results from one FSP sample prior to forming and the base material are shown in Figure 4.5. As can be seen, the results illustrate a significant local improvement in the ductility following FSP as compared with the base material. Also, there is a slight reduction in the tensile strength, but results still exceed that of the base material minimum specification, as specified by the Aluminum Association (Aluminum

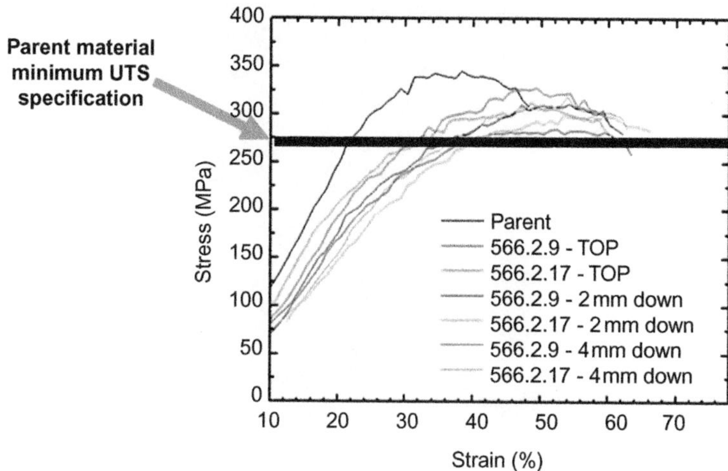


*Figure 4.5 Stress–strain behavior of FSP samples prior to forming and the base material. Large elastic range is due to the use of mini tensile testing equipment and should be ignored as artifact for the discussion of specimens.*

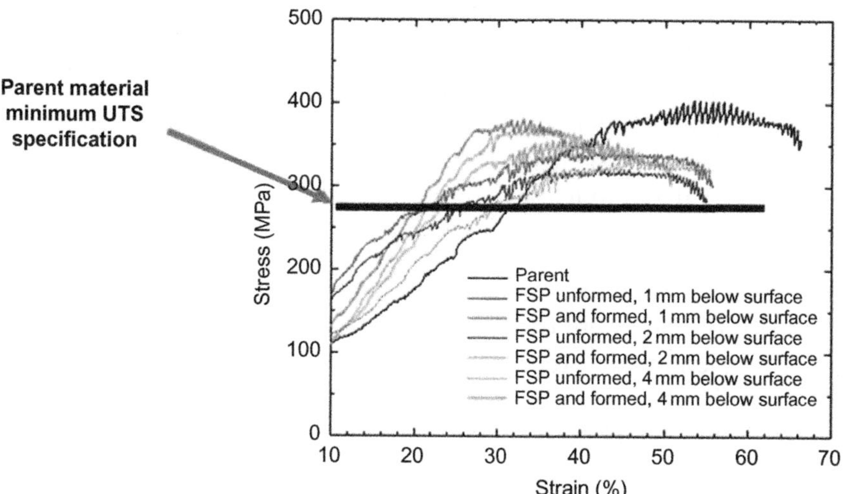


*Figure 4.6 Tensile test results of formed FSP samples versus location.*

Standards and Data, 2009). Elongation within the friction stir processed region was measured to be approximately 30%.

Mini tensile testing was also performed after forming. The test results for postformed samples, in the same locations as the unformed sections, are shown in Figure 4.6. In this case, the tensile strength has increased

*Figure 4.7 Photograph of a 6-m-long strip demonstrating the feasibility of using FSP technology for enhanced formability.*

and the ductility has decreased as compared to the unformed material. For example, at a depth of 1 mm below the surface in tension, the ultimate tensile strength (UTS) has increased from ~330 MPa following FSP to ~370 MPa following FSP plus bending. Strain hardening appears to increase the UTS to a depth of at least 4 mm. This strength increase is not unexpected, given the strain hardening introduced by the forming operation. The tensile strength after forming is near that of the parent material and well above the parent material minimum.

The last step was to attempt to form multiple full-size prototype components. Using the same parameters previously developed, 6-m-long strips of plate were friction stir processed along the intended bend line. The friction stir processing starts and stops were then removed. Subsequently, the parts were successfully formed in a large press brake. The resulting formed parts are shown in Figure 4.7, demonstrating the feasibility of using the FSP technology to locally enhance the ductility of aluminum and employ the process to allow for forming of angles and other potential shapes.

## 4.3 CASE STUDY DESCRIPTION

The feasibility trials were performed with consultation of the Navy and its shipyards. Following completion of the 6 m formed prototype parts, final results were presented to the Navy and its shipyards. Based on success of these trials, the team was asked to seek NAVSEA approval of the concept and American Bureau of Ships (ABS) certification for production. It is noted that the NAVSEA approval of the concept is different from the ABS certification for production. As will be discussed in the following sections, the NAVSEA approval of the concept involves a large suite of testing ranging from static testing to dynamic testing to corrosion testing on a single thickness. The ABS certification process is performed on all thicknesses and is used to qualify a friction stir processing process for production. The latter involves a smaller suite of tests, typically limited to static destructive testing and nondestructive examination (NDE) per existing friction stir welding qualification specifications. Only the testing associated with the NAVSEA approval process is the subject of this book, though subsets of the data are applicable to ABS certification.

Upon the request from Marinette Marine to seek NAVSEA approval of the use of friction stir processing for forming of aluminum for marine structural angles, girders, and frames, a summary report of the feasibility work was written and provided to the US Navy. As part of the report, a production approval of the concept was requested. As part of the response, the United States provided a NAVSEA technical publication/specification, "Material Systems Requirements" (T9074-AX-GIB-010/100) indicating requirements for approval of a new concept (NAVSEA #T9074-AX-GIB-010/100, 1999). This specification and its requirements formed the basis for the case study of the subject of this publication.

Since the proposed concept involved a new fabrication process that changes the microstructure, a number of material and structural properties were required to be established as required by T9074-AX-GIB-010/100, prior to using FSP/bending in production. This required information (material and structural properties) was to be proposed by Friction Stir Link and its partners in a Material Systems Information (MSI) document (Smith et al., 2009), which was the foundation for the case study. The MSI document/case study description was generated and agreed upon with NAVSEA prior to initiating any material property or structural property testing. Given the requirements of the NAVSEA qualification requirements document, the testing was likely to involve the most expansive comparison between friction stir processing (and welding), gas metal arc welding, and the base material properties versus any known publication to date.

Since there was near term interest in implementation of the friction stir processed alternative approach, the case study was developed with this consideration. This had two significant implications:

1. An approach was taken to limit the approval to a subset of potential applications. This is referred to as the initial approval. Additional testing was to be performed in the future to expand the approval to include additional application. The initial approval was limited to certain applications, with the following conditions:
   a. Aluminum alloy AA 5083-H116 in 8 mm (5/16″) thickness.
   b. Single pass friction stir processing. This limits thickness to approximately ½″ (12 mm) or less. Initial approval would be for material thicknesses ranging from 4 to 13 mm for Navy applications.
   c. Internal applications only, except for one external test case in order to develop some natural exposure corrosion data. The majority of applications are applied internally to the ship superstructure.
   d. Any structural angles longer than the length of the plate would still be fabricated with the traditional GMAW approach, though the long-term objective will be obtained approval to use friction stir welding to create the splices in the flat condition, prior to forming.
2. The proposed test methods were categorized into four different areas:
   a. Category 1: Information already on hand. That is data exists from feasibility trials conducted during the initial feasibility state.
   b. Category 2: Additional information (not already in hand) but needed for approval prior to production. That is, data did not exist and was required prior to production for the subset of applications noted above.
   c. Category 3: Additional information needed for approval but will be provided at a later date (not prior to production). Technical justification was provided to explain why there is a high probability that the remaining tests results would not have an adverse impact to the performance of the component. Testing was to be performed, following completion of the Category 2 testing.

**Table 4.1 Summary of Different Sample Types**

| Type Designation | Description | Notes |
|---|---|---|
| A | Base material[a] | Testing performed both perpendicular and parallel to the rolling direction where appropriate |
| B | FSP of flat plate (no forming)[a] | FSP parallel to the rolling direction |
| C | FSW splice of flat plate (no forming)[a] | FSW perpendicular to the rolling direction |
| D | FSP of flat plate + forming at small radius[b] | FSP and bending parallel to the rolling direction |
| E | GMAW of flat plate (no forming)[a] | GMAW perpendicular to the rolling direction |
| F | Base material formed at large radius[a] | Bending perpendicular to the rolling direction |
| G | GMAW of formed base material[c] | GMAW perpendicular to the rolling direction |
| H | FSW splice flat plate + FSP flat late + forming at small radius[a] | FSW perpendicular to the rolling direction and FSP and bending parallel to the rolling direction |

[a]*Samples created in test coupon size only (12").*
[b]*Samples created in test coupon size (12") and full-length samples (18 ft).*
[c]*Samples created in full-length samples only (18 ft).*

d. Category 4: Additional data that was not required for approval but should have to demonstrate the process/technology. That is data was not required prior to production and technical justification was not required. Testing would still be performed in the future, following completion of the Category 2 and 3 testing.

With these considerations a test matrix was generated with various sample types and testing methods. The final objective was to be able to compare material and structural property data of the proposed approach versus the traditional approach (gas metal arc welded assemblies). The test matrix included eight different sample types as indicated in Table 4.1. The test sample types considered all of the initial approval application characteristics as noted in 1) above, (e.g., AA 5083-H116), with the exception being inclusion of FSW splices for eventual approval of this configuration for structural angles longer than the length of plate. The FSW spliced configuration was included in the case study for several reasons.

1. This was the most common additional characteristic or requirement of other structural angles not included in the list above.

2. There was limited published technical data on response of FSW to the various test methods. This information will be valuable to the general use of FSW in the marine industry.
3. Most of the inferences for technical justification (Category 3) were made from FSW data, not FSP. Though FSW is a very similar process, confirmation of similar technical capability or results were important.

As a part of the case study, multiple test coupons were made for all sample types indicated in Table 4.1. All samples were created on 8-mm-thick (5/16″) material. The samples were all labeled with the sample type designation indicated in Table 4.1 and then the test method indicated in Table 4.2. In all, there were 19 different test methods (include spare samples) as indicated in Table 4.2. The testing category is also indicated in the middle column per the description above. There were also multiple samples of most test types for repeatability purposes, with the number of samples indicated in Table 4.2. Lastly, any special notes and the specification to which the testing was performed are indicated in the rightmost column. This test matrix was included in the MSI document/case study plan.

In the case study, all test methods used the small test coupon (12″ length) except for those noted in the rightmost column, for example, distortion measurements. It is also noted that not all sample types were tested with all methods, since this was not practical. For example, formed samples were not subjected to macro transverse tensile testing.

Although not indicated in the test matrix, a portion of the samples were also subject to NDE. This included the macro transverse tensile test samples and all of the full-length formed samples. The NDE included 100% visual and 100% radiography prior to bending and 100% etched dye-penetrant inspection after bending. Acceptance criteria for qualification was to MIL-STD-2035, Class I with the added visual testing criteria that appear in NAVSEA Project Peculiar Document (PPD) #802-7651533A to inspect for visual aspects unique to FSW/FSP (exit hole, surface tool markings, etc.) (NAVSEA #802-7651533A, 2006a).

## 4.4 INITIAL COMPARATIVE PROCESS QUALIFICATION

Prior to initiating the case study and applying any specific test method to any one sample type, an initial process qualification was performed

**Table 4.2 Summary of Different Tests Used in This Study**

| Test Method Designation | Test Method or Material Property | Test Cat. | No. of Samples | Specification and Other Notes |
|---|---|---|---|---|
| A | Macro cross section | 2 | 1 | ASTM E 340-00. Note: ASTM specification requires same preparation procedures as MIL-STD248 but specifies a particular etchant, whereas MIL-STD248 does not specify the etchant. |
| B | Microhardness | 2 | 1 | ASTM E384-09. Note: MIL-STD248 does not go in the details of microhardness testing. Section 4.4.2.2 of MIL-STD248 mentions Table VII as a guide. Table VII only specifies the dimensions of the hard-faced specimens. |
| C | Macro transverse tensile test | 2 | 3 | AWS B4.0 with 5 to 1 width to thickness ratio. Includes UTS, yield strength (YS), percent elongation, and reduction of area (R of A). |
| D | Young's modulus | 2 | 3 | ASTM E756 |
| E | Shear modulus | 2 | 3 | ASTM E756 |
| F | Mini transverse tensile test | 1,2 | 3 | Includes UTS, YS, percent elongation, and R of A |
| G | Mini longitudinal tensile test | 2 | 3 | Only at location of minimum transverse UTS. Includes UTS, YS, percent elongation, and R of A. See Section 3.6 for test locations and description test coupon geometry. |
| H | Distortion measurements | 2 | 3 | Full-length samples only |
| I | Buckling test | 3 | 3 | Subsection of full-length samples |
| J | Tripping and column test | 3 | 3 | Subsection of full-length samples |
| K | SCC | 2 External, 3 internal | 5 | ASTM G44 |
| L | IGC | 2 External, 3 internal | 5 | ASTM G67 |
| M | Pitting corrosion | 2 External, 3 internal | 5 | ASTM G46 |
| N | Crevice corrosion | 2 External, 3 internal | 5 | Consult ASTM G48 and G78 |
| O | Natural exposure | 3 | 5 | Single application to be chosen by Marinette Marine. |
| P | Corrosion resistance | 2 External, 3 internal | 5 | ASTM B117 |

(*Continued*)

**Table 4.2 (Continued)**

| Test Method Designation | Test Method or Material Property | Test Cat. | No. of Samples | Specification and Other Notes |
|---|---|---|---|---|
| Q | Fatigue | 3 | 10<br>15 (mini) | ASTM E466-07, mini fatigue of parent material, FSP, and FSP + bend specimens. Both residual stresses and distortion will influence fatigue life. We are planning to do regular specimen (3.5 mm thick) and mini fatigue (1 mm thick). When we extract specimens, the level of distortion and residual stress effect changes. So, the full effect cannot be captured by coupon specimens. |
| R | Corrosion fatigue | 3 | 10 | ASTM E466-07 and ASTM F1801-97 |
| S | Exfoliation corrosion | 2 External, 3 internal | 5 | ASTM G66 |
| T | Spares | N/A | 5 | |

for the three critical processes used in the traditional (gas metal arc welding) and the proposed approaches (friction stir processing and friction stir welding). These qualifications were performed to ensure that the welding or processing parameters used to create the samples were acceptable and meet applicable industry standards. In this initial qualification, these three processes (FSP, FSW, and GMAW) were also compared against the base metal. The qualification for the three processes included basic testing processes, such as tensile testing, bend testing, and macro cross sectioning to ensure that the base line processes for each of the manufacturing processes was acceptable. As noted, in the traditional approach, GMAW is used to splice the various shorter sections together to create a longer section. With the proposed approach, FSP is used along the future bend line to enhance the ductility of the material locally, and FSW can be used to join longer sections versus GMAW.

Destructive testing included tensile testing, bend testing, and macro cross sectioning. Tensile testing was performed per AWS B4.0 (2007) with minimum thickness to gage width ratio of 5 to 1 using strength specification as indicated in AWS D1.2 (2008) and bend testing (root and face) diameter per AWS B4.0. In addition, all the processed samples were cross sectioned. For the initial qualification, the minimum tensile strength was to be 275 MPa (40 ksi) per AWS D1.2 and ABS Requirements for Materials and Welding, Part 2, Chapter 5, Appendix 1 and the bend radius was 50 mm (2″). The welds were also nondestructively examined. Specific procedures included etched dye-penetrant testing, conventional ultrasonic testing, and visual inspection. The testing procedures were performed per the NAVSEA PPD #802-7651532A (2006b). Although this NAVSEA specification was developed specifically for procedure qualification for FSW, it was also selected for FSP for consistency. This decision was made for several reasons. First there is significant similarities between friction stir welding and friction stir processing. Additionally, production processes had been qualified to the specification for other Navy applications. Lastly, there is a general lack of other specifications with their being none available for FSP and limited specifications applicable to FSW.

## 4.5 GMAW QUALIFICATION AND DESTRUCTIVE TESTING

For the traditional approach (GMAW), Navy ship fabricators were consulted for specific welding parameters and setup that were used for fabrication of structural angles. The parameters were replicated by Friction Stir Link, Inc. technicians and implemented to join 8-mm-thick (5/16″) AA 5083-H116 flat plate samples for the initial process qualification. The qualification plates were prechamfered with a 60° included angle to a depth of 5 mm on one side. After chamfering, the plates were wire brushed and then wiped down with alcohol to remove contaminants. They were then manually welded on the chamfered side with a pulse GMAW process using argon shielding gas, 0.035″ diameter AA 5356 filler wire, with a travel speed of approximately 8 mm/s (20 IPM), amperage range of 52–54 amps, and wire feed speed of approximately 300 mm/s (720 IPM) per the weld parameter specification (WPS). Next, the plates were flipped over and a shallower back gouge performed to a depth of about 4 mm. The plates were then wire brushed and wiped down with alcohol before the final manual weld was made with similar settings as the first weld. After welding, the plates were ground to remove any weld reinforcement. A portion of a gas metal arc welded sample test plate after grinding is shown in Figure 4.8.

Subsequently, the welds were destructively and nondestructively examined. The destructive examination included tensile and bend

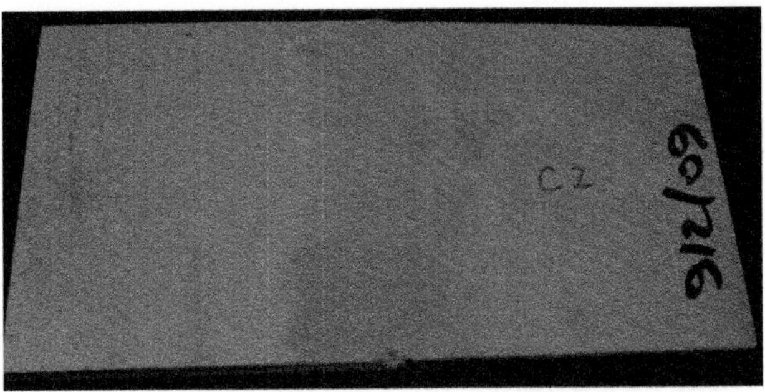


*Figure 4.8 Photograph of a portion of a gas metal arc welded sample test plate after grinding.*

testing per the specifications noted previously. The nondestructive testing included visual inspection and dye-penetrant examination. With the destructive and nondestructive inspection data, a procedure qualification record (PQR) was created to document the results. This is shown in Figure 4.9 indicating that the welds created using the WPS were acceptable. In the actual application (splicing of the sections), there is also a vertical up weld that is required. The setup for this welding is shown Figure 4.10. Due to the increase in difficulty for the vertical up weld, the qualification testing as noted above was repeated to ensure the process could be used and replicated for the vertical up portion of the weld.

During the process qualification phase, the relative sensitivity or difficulty of the GMAW process versus friction stir processing were readily evident. For example, for GMAW, the precleaning operation was critical and required extra care to ensure that hydrocarbons and other contaminants were completely removed. The plates that were welded were required to be ground, then wiped carefully with alcohol. Otherwise, porosity would develop within the weld, reducing its relative quality and its strength. A similar level of cleanliness is not required for FSP. Although FSP samples used for the subject of this document were ground prior to FSP, there has been no evidence this is required with material delivered and handled with typical production shop practices (no water stains or heavy oils). No reductions in strength or quality have been observed when the parts are not ground or wiped with alcohol. Clearly, there is a level at which contamination will affect FSP results, but no systematic study has been performed. However, it is clear that the relative sensitivity of FSP to contaminants is dramatically less than that of GMAW.

## Procedure Qualification Record (PQR)

Friction Stir Link Inc.

Process: Gas Metal Arc Welding
PQR # & Date: 725A.C.PQR – 12-3-09
Testing Date: 6-Oct-09
Part Number: 725.A.C.1-3

☐ Periodic Quality Testing ☒ Initial Procedure Qualification
WPS Reference #: 725.E.C GMAWWPS 091006
Date Product Completed: 6-Oct-09

*Minimum tensile strength requirement per AWS D1.2 GMAW Process* — 40 KSI (275 Mpa)

***Tensile Test*** — *AWS B4.0 Standard Method mechanical Testing of Welds*

| Weld Identification Number | Test Type TW=Transverse LW=Longitudinal | Location S=Start M=Middle E=End | Sample Width (mm) | Sample Thickness (mm) | Failure Load (N) | Tensile Strength (MPa) | Yield Strength (Mpa) | Elongation (%) | Fracture Location | Results Accept / Reject |
|---|---|---|---|---|---|---|---|---|---|---|
| 725.E.C.1 | LW | S | 38.303 | 8.077 | 84332 | 272.6 | 142.7 | 12 | HAZ | |
| 725.E.C.2 | LW | M | 37.821 | 7.976 | 77946 | 258.4 | 160.0 | 10 | HAZ | |
| 725.E.C.3 | LW | E | 37.668 | 8.052 | 81253 | 267.9 | 156.5 | 11 | HAZ | |
| | | | | | | | — | — | | |
| | | | | | | | — | — | | |
| | | | | | | | | | | |

*Notes: Fracture Location BM=Base Mat'l HAZ=Heat Affected Zone TW=Through Weld RS=Retreating Side AD=Advancing Side*

***Bend Test -*** — *FSL Standard Test Method for Bend Testing (WI-QA-002)*

| Weld Identification Number | Location | Type of Bend Test | Former Diameter | Results (Accept / Reject) |
|---|---|---|---|---|
| 725.E.C.1 | S | Root / (Face) / Side / Longitudinal | 50.8 (2") | ACCEPT |
| 725.E.C.2 | M | Root / (Face) / Side / Longitudinal | 50.8 (2") | ACCEPT |
| 725.E.C.3 | E | Root / (Face) / Side / Longitudinal | 50.8 (2") | ACCEPT |
| | | | | |

***Visual Examination -***
*See FSL Visual Inspection Procedure NAVSEA No. 802-7651533A*

| Weld Identification Number | Results (Accept / Reject) |
|---|---|
| 725.E.C.1 | ACCEPT |
| 725.E.C.2 | ACCEPT |
| 725.E.C.3 | ACCEPT |
| | |

***Penetrant Examination -***
*See FSL Penetrant Inspection Procedure NAVSEA No. 802-7651533A*

| Weld Identification Number | Results (Accept / Reject) |
|---|---|
| 725.E.C.1 | ACCEPT |
| 725.E.C.2 | ACCEPT |
| 725.E.C.3 | ACCEPT |
| | |

***Ultrasonic Examination -***
*See FSL Ultrasonic Inspection Procedure NAVSEA No. 802-7651533A*

| Weld Identification Number | Results (Accept / Reject) |
|---|---|
| 725.E.C.1 | ACCEPT |
| 725.E.C.2 | ACCEPT |
| 725.E.C.3 | ACCEPT |

***Macroetch Examination -***
*See FSL Marcoetch Inspection Procedure WI-QA-004*

| Weld Identification Number | Results (Accept / Reject) |
|---|---|
| 725.E.C.1 | ACCEPT |
| 725.E.C.2 | ACCEPT |
| 725.E.C.3 | ACCEPT |

***Additional Information / Comments -***

*Figure 4.9 Illustration of a PQR that was used for gas metal arc welding.*

*Figure 4.10 The setup used for gas metal arc welding in this study.*

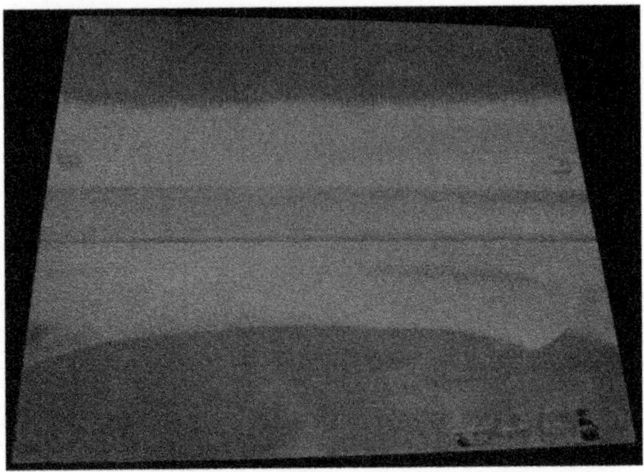

*Figure 4.11 A sample of an FSP initial qualification plate.*

## 4.6 FSP QUALIFICATION AND DESTRUCTIVE TESTING

For the initial qualification of FSP, a similar approach was taken as described above. The differences included the setup for FSP and parameters peculiar to FSP. In addition, the inspection was altered due to the specific differences of FSP versus GMAW. These inspection differences or exceptions were performed as noted in the referenced NAVSEA specifications. The FSP parameters used were based on lessons learned from the initial feasibility trials. To summarize, the travel speed was 6 mm/s, the rotation speed 500 RPM, and an FSP tool that enabled FSP depths of approximately 2 mm. The specific details of the process were recorded in a WPS. It should be noted that the WPS format is altered for consideration of friction stir processing but was in a format that meets NAVSEA #802-7651532A. A sample of an FSP initial qualification plate is shown in Figure 4.11. As with the GMAW samples, the FSP samples were destructively and nondestructively tested as previously described and per the previously indicated specifications. The PQR resulting from this test analysis is shown in Figure 4.12.

## 4.7 FSW INITIAL QUALIFICATION AND DESTRUCTIVE TESTING

For the initial qualification of FSW, a similar approach was taken to what was performed with the FSP. The differences mostly included the setup for FSW, as FSW requires creation of weld joint and fixturing to secure the two plates appropriately. The inspection was the same approach as with FSP. The FSW parameters used were based on

**Procedure Qualification Record (PQR)** — Friction Stir Link

Process: Friction Stir Welding
PQR # & Date: 725.B.C – 10-5-09
Testing Date:
Part Number: 725.8MM.FSP.1-3

☐ Periodic Quality Testing ☒ Initial Procedure Qualification
WPS Reference #: 725.B.C F110 091005
Date Product Completed: 2-Oct-09

*Minimum tensile strength requirement per AWS D1.2 GMAW Process* — 42 KSI (285 Mpa)

*Tensile Test* — *AWS B4.0 Standard Method mechanical Testing of Welds*

| Weld Identification Number | Test Type (TW=Transverse, LW=Longitudinal) | Location (S=Start, M=Middle, E=End) | Sample Width (mm) | Sample Thickness (mm) | Failure Load (N) | Tensile Strength (MPa) | Yield Strength (Mpa) | Elongation (%) | Fracture Location | Results Accept / Reject |
|---|---|---|---|---|---|---|---|---|---|---|
| 725.8MM.FSP.T1 | LW | S | 23.927 | 7.798 | 62598 | 335.5 | N/A | N/A | HAZ | Accept |
| 725.8MM.FSP.T2 | LW | M | 21.996 | 7.849 | 58059 | 336.3 | N/A | N/A | HAZ | Accept |
| 725.8MM.FSP.T3 | LW | E | 17.932 | 7.874 | 47757 | 338.2 | N/A | N/A | HAZ | Accept |
| | | | | | | | ----- | ----- | | |
| | | | | | | | ----- | ----- | | |

*Notes: Fracture Location BM=Base Mat'l HAZ=Heat Affected Zone TW=Through Weld RS=Retreating Side AD=Advancing Side*

*Bend Test -* — *FSL Standard Test Method for Bend Testing (WI-QA-002)*

| Weld Identification Number | Location | Type of Bend Test | Former Diameter | Results (Accept / Reject) |
|---|---|---|---|---|
| 725.8MM.FSP.B1 | S | Root / (Face) / Side / Longitudinal | 12.7mm (.5") | Accept |
| 725.8MM.FSP.B2 | M | Root / (Face) / Side / Longitudinal | 12.7mm (.5") | Accept |
| 725.8MM.FSP.B3 | E | Root / (Face) / Side / Longitudinal | 12.7mm (.5") | Accept |
| | S | Root / Face / Side / Longitudinal | 12.7mm (.5") | Accept |
| | M | Root / Face / Side / Longitudinal | 12.7mm (.5") | Accept |
| | E | Root / Face / Side / Longitudinal | 12.7mm (.5") | Accept |

*Visual Examination -* — *See FSL Visual Inspection Procedure FSL-489-04*

| Weld Identification Number | Results (Accept / Reject) |
|---|---|
| | Accept |
| | Accept |
| | |
| | |

*Penetrant Examination -* — *See FSL Penetrant Inspection Procedure FSL-489-03*

| Weld Identification Number | Results (Accept / Reject) |
|---|---|
| 0 | Accept |
| 0 | Accept |
| | |
| | |

*Ultrasonic Examination -* — *See FSL Ultrasonic Inspection Procedure FSL-489-002*

| Weld Identification Number | Results (Accept / Reject) |
|---|---|
| 0 | Accept |
| 0 | Accept |
| | |

*Macroetch Examination -* — *See FSL Marcoetch Inspection Procedure WI-QA-004*

| Weld Identification Number | Results (Accept / Reject) |
|---|---|
| | Accept |
| | Accept |
| | Accept |

*Figure 4.12 Illustration of a PQR that was used for friction stir processing.*

lessons learned from the initial feasibility trials and an FSW tool was used that enabled full penetration depth welds. The specific details of the process were recorded in a WPS. As with the FSP, the WPS for FSW was in a format that meets NAVSEA #802-7651532A. As with the GMAW and FSP samples, the FSW samples were destructively and nondestructively tested as previously described and per the previously indicated specifications.

## 4.8 MACRO CROSS SECTION COMPARISON

### 4.8.1 Introduction

Macro cross sections were created for all sample types to investigate the macrostructure of the different zones including that of the base metal, friction stir processing, and gas metal arc welding. Both formed

*Figure 4.13 A cross section of a gas metal arc weld created in the flat position.*

and unformed samples were cross sectioned where appropriate. This cross sectioning was performed to ASTM E 340 (2000). In all cases, the samples were etched with 5% hydrofluoric acid after polishing the surface with 1 μm diamond paste as the last polishing step.

### 4.8.2 Macro Cross Section Inspection Results

Figure 4.13 shows a cross section of a gas metal arc weld created in the flat position (1G). This weld was fabricated per the gas metal arc WPS developed in the initial process qualification work. After welding, the weld reinforcement was removed via abrasive grinding. As is seen, the weld is a two-pass weld, with one weld performed after a prechamfering operation. The initial chamfer is 5 mm in depth with a 60° included angle. After the first weld was created, the weld reinforcement was ground off. Then the plate was flipped over and the back side was gouged to a depth of 4 mm per the WPS. The back side was then wire brushed and wiped down with alcohol before the second weld was created in the flat position (1G position). As with the top side, the weld reinforcement was removed via an abrasive grinding operation. After completion of the weld process, a sample was processed for cross sectioning. The resulting cross section reveals no unusual features, but minor amounts of porosity can be seen (areas of white). However, these are within allowable limits as noted in AWS D1.2 (2008): Structural Welding Code—Aluminum. In addition, the welds were nondestructively inspected per references sited in the NAVSEA specification. Of note is the cast microstructure (dark areas) versus the wrought microstructure of the base metal away from the weld. The precleaning operation had a significant impact on the level of porosity and subsequent strength of the weld as will be discussed in the following sections.

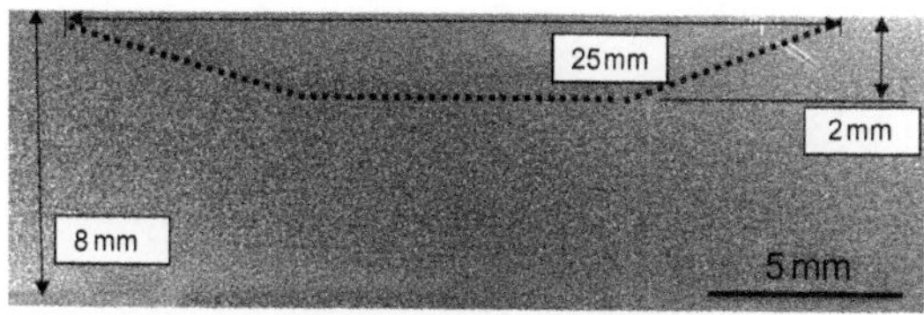


*Figure 4.14 A cross section of the friction stir processed plate. The FSP run was parallel to the rolling direction.*

*Figure 4.15 A cross section of the friction stir processed region after forming to an 8 mm (5/16") internal radius.*

Figure 4.14 shows a cross section of the friction stir processed plate, with FSP parallel to the rolling direction. FSP was performed with the WPS developed in the initial qualification trials. The FSP zone (stir zone) is relatively difficult to discern. Given this, the FSP zone is outlined with the dotted line in the photograph. There are several reasons for the difficulty in discerning or being able to highlight the stir zone. First, FSP is an autogenous process (there is no introduction of filler metal of a different alloy). Second, there is limited impact of the applied heat, yielding a limited heat-affected zone (HAZ), due to the lower heat input of the FSP process. Finally, the alloy used in this study was a nonheat treatable alloy, which is less impacted by the introduction of heat. The FSP zone is approximately trapezoidal in shape, being about 2 mm deep in the center, as noted by the whiter area and darker areas. The overall width of the FSP zone is about 25 mm. A finer grained microstructure is evident in the FSP stir zone.

Figure 4.15 shows a cross section of the friction stir processed region after forming to an 8 mm (5/16″) internal radius. The bend axis was parallel to the rolling direction and centered about the FSP zone. However, of important note is the lack of any visible

*Figure 4.16 A macro cross section of the base metal.*

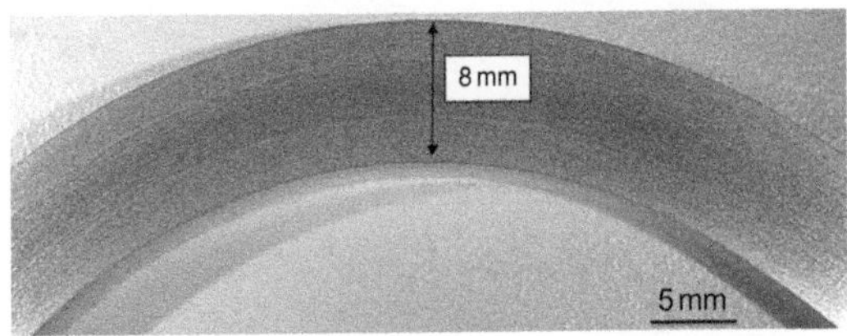


*Figure 4.17 A macro cross section of the formed base metal.*

cracking, indicating the ability of FSP to locally enhance ductility. Little, if any, difference is noted in the macro cross section other than the obvious fact of the overall shape of the macro cross section.

FSW was performed with the WPS developed during the initial qualification trials. As with the FSP, the FSW stir zone is comparatively difficult to etch and visualize and again is outlined with the dotted line in the photograph. The FSW stir zone is approximately trapezoidal in shape, but extends to the bottom of the base metal, as the FSW is required to have full penetration. This can be approximated by the whiter area versus darker areas. The overall width of the FSW stir zone on the face surface is about 25 mm and at the root surface it is approximately 10 mm.

Figure 4.16 shows a macro cross section of the base metal. The view is transverse to the rolling direction. The cross section shows a uniform structure. The grain orientation is also observable.

Figure 4.17 shows a macro cross section of the formed base metal. The forming has been performed with the bend axis perpendicular to the rolling direction (direction of higher elongation) to an internal bend radius of 1.25″. This is the bend radius used for the spliced section using the traditional approach. The cross section shows a uniform structure as would be expected.

### 4.8.3 Summary

A review of the macro cross sectioning results highlights the differences between gas metal arc welding, friction stir processing, friction stir welding, and the base metal. Macro cross sections of both unformed and formed samples were reviewed and discussed. It was shown that FSP and FSW have little impact on the material, as is evidenced by the relative inability to visualize the FSP and FSW stir zone versus the base metal. The fine grain microstructure of the resulting FSW and FSP is evident in the cross section. To the contrary, the gas metal arc welding is readily visible, mostly due to the introduction of the filler metal which is a different alloy and the fact that the fusing welding process yields a cast microstructure in the weld nugget. In addition, the relative difficulty to perform GMAW versus FSP is seen through the differences in the welding cross section results where some porosity was evident in GMAW, despite thorough precleaning procedures being implemented.

## 4.9 MICROHARDNESS TEST RESULTS AND COMPARISON

### 4.9.1 Introduction

Microhardness measurements are reviewed for the various sample types, with microhardness measurements completed per ASTM E 340-00 (2009). Microhardness measurements are reviewed for all sample types, including the base material, friction stir processed, friction stir welded, and gas metal arc welded samples. In addition data from both the unformed and formed conditions are summarized where appropriate to the application. Microhardness data is from multiple depths from the surface, and where appropriate, a 2D grid of data points was generated and reviewed.

### 4.9.2 Results

In Figure 4.18 the through thickness microhardness of the base metal is shown at various depths. For this figure, the microhardness was measured at 0.5 mm increments through the thickness of the material. The hardness is highest near the surface and lowest in the center, ranging from about 85–95 VHN. Higher hardness readings near the surface are typically expected for rolled or work-hardened plate, since the

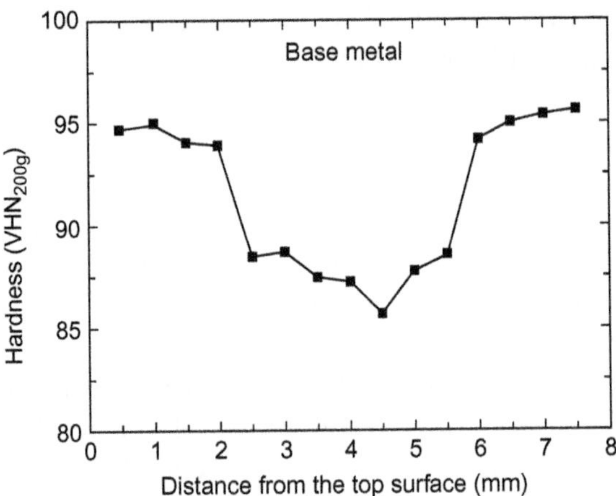


*Figure 4.18 Through thickness microhardness of the base metal at various depths.*

work hardening from the rolling operation has the most effect on near surface material.

In Figure 4.19, the microhardness of the base metal formed to the 1.25″ internal radius is shown at various depths. The 1.25″ radius is the internal dimension that is used for the formed structural angle with the traditional fabrication approach. The data are taken at 1 mm increments across the formed zone. The hardness is highest near the surface and lowest in the center, ranging from about 95–115 VHN. The data demonstrate the work hardening that occurs during the forming process, as the hardness is 20–25 VHN higher following forming.

In Figure 4.20, a microhardness map of a gas metal arc weld sample is shown. This is for the weld created in the flat position (1G position). The map is oriented such that the first weld is at the top of the map. The data were generated at 0.5 mm increments in the vertical direction through the thickness of the material and in 0.5 mm increments from 17 mm to the left and right of the weld centerline. The hardness is highest far away from the weld zone and outside the relatively large HAZ, approaching values similar to the base material. The data demonstrate that the minimum hardness is in the HAZ, approximately 70 VHN. There are lower values indicated in the weld zone, but it is believed these measurements are being affected by local small amounts of porosity. Overall, the data indicate the significant effect of the applied heat of the gas metal arc welding process. Importantly,

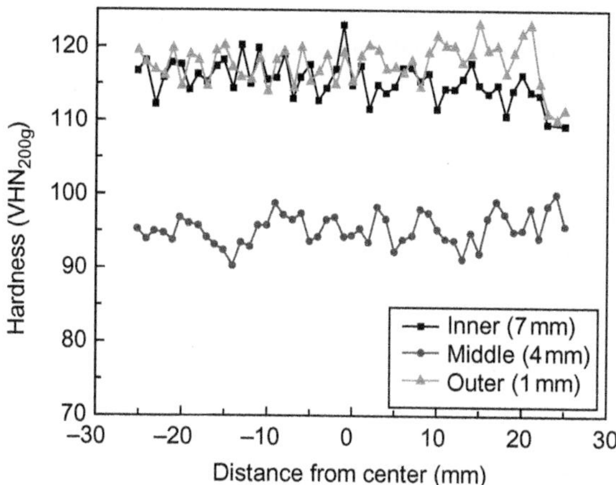


*Figure 4.19 Microhardness of the base metal formed to the 1.25" internal radius at various depths.*

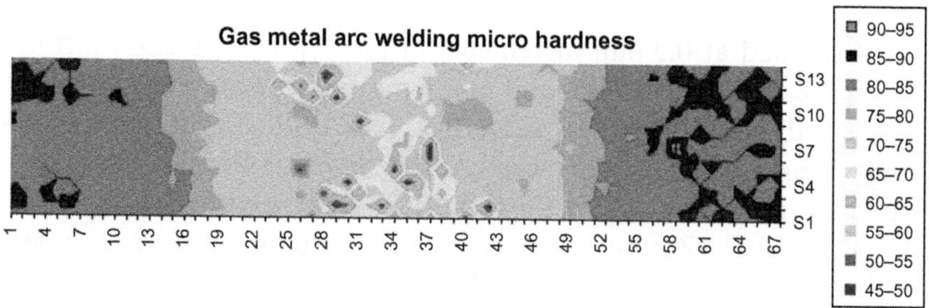


*Figure 4.20 A microhardness map of a gas metal arc weld sample. Regions of low hardness in the weld pool indicates micro defects.*

these values are what exist in the final spliced product, as there are no forming operations or other operations that would increase the hardness above these measurements (see discussion of FSP and formed FSP samples for comparison).

In Figure 4.21, a microhardness map of the vertical up (3G position) gas metal arc welding sample is shown. The hardness map is oriented such that the first weld is at the top of the map. The data are taken at 0.5 mm increments in the vertical direction through the thickness of the material and from 0.5 mm increments from 20 mm to the left and right of the weld centerline. Results are similar to the horizontal weld but indicate a narrower HAZ. The exact source of the difference is unknown, but one potential reason includes variation in welding speed, given the manual nature of the process. Regardless, this

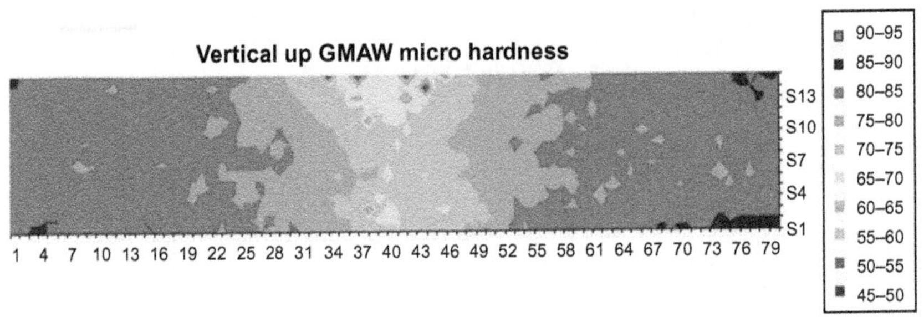


*Figure 4.21 A microhardness map of the vertical up (3G position) gas metal arc welding sample.*

highlights the variability of a manual versus a machine controlled process such as FSW or FSP.

In Figure 4.22 a microhardness map of a friction stir processed sample is shown. These data are for an unformed sample. The data were generated at 0.5 mm increments in the vertical direction through the thickness of the material and in 0.5 mm increments from 20 mm to the left and right of the FSP centerline. The FSP stir zone is approximately 2 mm deep per the macro cross section results. The hardness is highest far away from the weld zone (left, right, and below), and outside the small HAZ with values similar to the base material. The data also indicated that the minimum hardness is in the FSP zone or the HAZ and is approximately 75–80 VHN. Overall, the data indicate that the FSP process has less of an effect than the gas metal arc welding process, with Vickers hardness being 5–10 VHN higher than the GMAW samples. In addition a significantly smaller HAZ is observed.

Although an improvement over the GMAW, the hardness values of the FSP samples without forming will not be that which exists in the final product, as the subsequent forming operation will provide a local increase in hardness via strain hardening. It can also be seen that the base metal shows an interesting trend of being slightly softer in the middle of the sheet thickness. Such a trend is expected in rolled plates because of processing strain gradient during rolling in the thickness direction and associated hardness change.

In Figure 4.23, the microhardness indent locations on a friction stir processed and subsequently formed sample are shown. These data are taken at 1, 4, and 7 mm depths, and at 0.5 mm increments in the other direction. In Figure 4.24, the microhardness data from an FSP sample

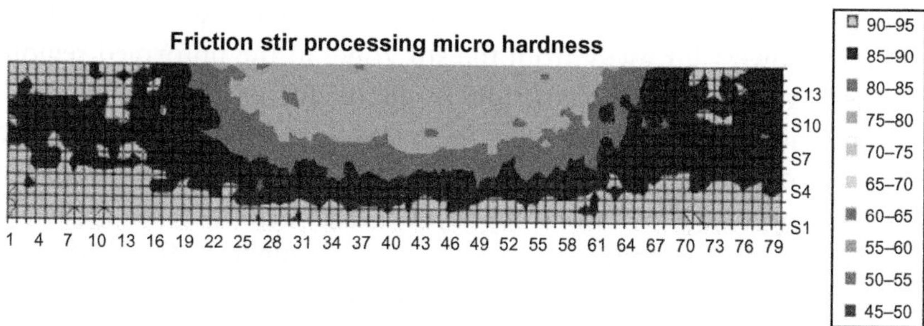


*Figure 4.22 A microhardness map of a friction stir processed sample.*

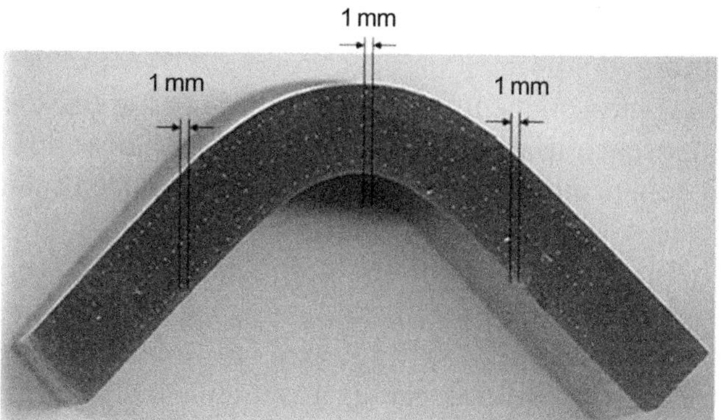


*Figure 4.23 The microhardness indent locations on a friction stir processed and subsequently formed sample.*

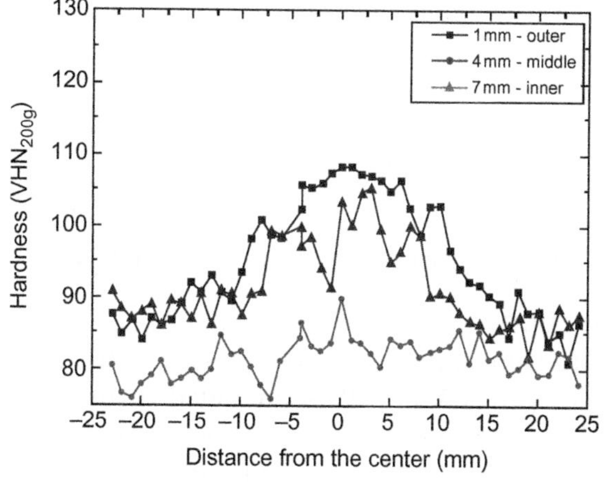


*Figure 4.24 The microhardness data from an FSP sample and then subsequently formed.*

and then subsequently formed is shown. These data now show that the hardness is lowest far away from the stir zone, in the unformed region. Additionally, the data demonstrates a significant increase in hardness over the unformed samples, with values approaching 110 VHN. The most significant increases are on the exterior tensile surface where the material is stretched or formed the most. These data indicate the hardness is even higher than the base metal throughout the FSP and formed region. Overall, the data indicate that FSP with subsequent forming has a significant benefit over the traditional approach. The traditional approach, employing gas metal arc welding, will only reduce hardness, while the FSP approach will result in microhardness in a range near the base material to in excess of the base material.

### 4.9.3 Summary

Microhardness measurements were taken through the thickness for the traditional approach using spliced gas metal arc weld sections, for friction stir processed material, for the friction stir welded material, and for base material, both in the unformed and formed conditions, where possible. The data demonstrated the negative impact of the gas metal arc welding process, where significant softening occurs. The data further demonstrated that FSP by itself leads to some softening, but the subsequent forming operation causes a subsequent significant increase in hardness. The forming operation yields final hardness values above the base material by up to 15 VHN on the Vicker's hardness scale. It is also noted that the reduction in hardness of the FSW is somewhat greater than that of the FSP, due to greater heat input resulting from the need to have a through thickness/full penetration joint. However, the reduction in hardness is not nearly as significant as that of the gas metal arc weld. As hardness tends to be correlated with reductions in strength, this would indicate that FSW should be a preferred splicing method over gas metal arc welding.

## 4.10 MACRO TRANSVERSE TENSILE TEST RESULTS

### 4.10.1 Introduction

A review of data from transverse tensile tests performed on a subset of the sample types per AWS B4.0 is presented in this chapter. Only a subset is possible to review, since the transverse tensile test requires a flat sample. Thus, formed samples could not be subjected to transverse tensile testing. The sample types reviewed included the base material, both parallel and transverse to the rolling direction, gas metal arc welded samples (horizontal and vertical up weld positions), friction stir processed samples, and friction stir welded samples. Data from a minimum of three samples from each type is summarized to allow for a measure of repeatability. UTS, yield strength, elongation, and reduction of area were measured for each sample type.

### 4.10.2 A Review of the Results

In Figure 4.25, UTS data are displayed for all the sample types tested. In the bar chart, both the average and the standard deviation are shown. The chart shows that the base material has fairly consistent UTS properties regardless of testing direction. The FSP causes a small reduction in UTS, as might be expected, due to the heat of the friction stir process which slightly anneals the material. The heat input in the FSP region shifts the microstructure from "H" temper toward "O" temper due to reduction in dislocation density. Concurrently, the grain size in FSP region refines due to recrystallization of microstructure. This correlates with the small reduction in hardness, which is a balance of the two types of microstructural changes. Gas metal arc welded samples have a larger reduction in strength versus the base material and the FSP samples. This can be expected due to the higher heat input of GMAW and the cast microstructure which results in very coarse grain size. The vertical up and horizontal position welds have similar strengths. The base material minimum specification is approximately 275 MPa and, as shown, the FSP samples have strengths significantly higher than this value.

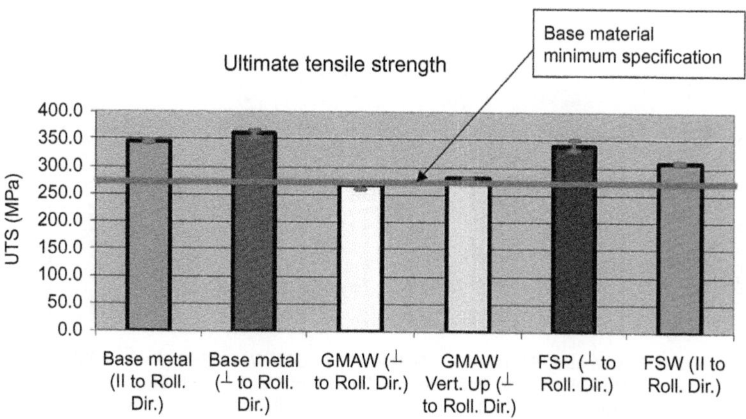


*Figure 4.25 A comparison of UTS for all the sample types tested.*

The yield stength data for the same sample types are shown in Figure 4.26. The data show similar trends as the UTS data, although reduction in yield strength versus the base material is more dramatic. For example, the gas metal arc welded samples show almost a 40% reduction in yield strength (approximately 250–150 MPa) versus only a 20% reduction for the friction stir processed samples. The postfriction stir processed samples have yield strengths of approximately 200 MPa. These data highlight a critical distinction between the two processes since the basis for many marine structural designs is the post gas metal arc welded strength. These data are only representative of the nonformed FSP samples. The post-FSP forming/bending causes work hardening which increases the strength (see Section 4.5) above the values shown in the figure. The nonbase material samples do have inhomogenous material properties, thus the yield strength and following data (elongation and reduction of area) are not a true measure of the local material properties. The mini tensile test results discussed in Section 4.5 will be more representative of local material properties.

Shown in Figure 4.27 are elongation data as measured and calculated from the transverse tensile tests. There are several important implications from these data. First, the base material has a significantly different elongation depending on the test direction. Transverse to the rolling direction, the material has a significantly higher elongation. This highlights the reason why the traditional process of cutting the plate in the width (short) direction, followed by forming

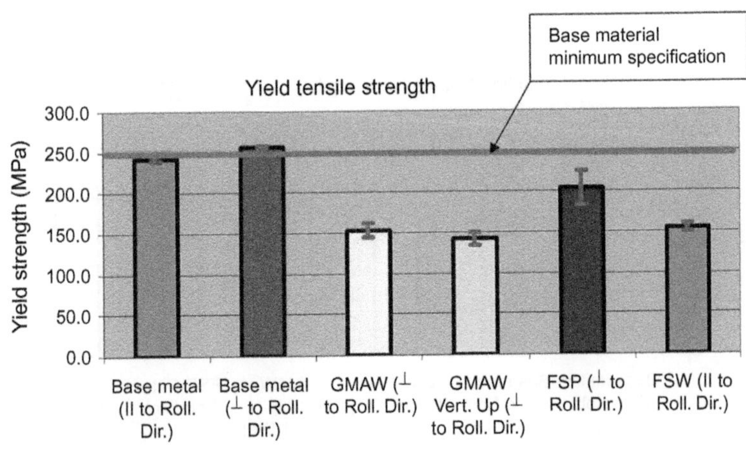


*Figure 4.26 A comparison of yield strength for all the sample types tested.*

perpendicular to the rolling direction, then splicing those short sections together, is performed in the manner that it is. The traditional production process avoids potential cracking caused by forming parallel to the rolling direction, but which could in theory avoid the splicing operation. However, because the material would likely crack, this alternative cannot be considered.

Next, the gas metal arc welded samples have the lowest elongation. This is due to the cast microstructure inherent with the melting and subsequent resolidification of the material that occurs due to GMAW. Lastly, the FSP sample has a significantly higher elongation than that of the base material tested perpendicular to the rolling direction. This highlights the capability of FSP to locally increase the ductility. However, the measured elongation does not exceed that of the base material tested parallel to the rolling direction. This could be considered contradictory to the results which show that FSP samples can be formed to a tighter radius than the base material. However, friction stir processing is only partial penetration, that is, only a couple of millimeters deep. This highlights the previously discussed issue of nonhomogenous material properties through the thickness and the effect on the data generated from a macro tensile test. That is, elongation of the microstructure on the tensile surface, where it is most needed, is likely considerably higher (locally) than the elongation of the composite microstructure and even the base alloy.

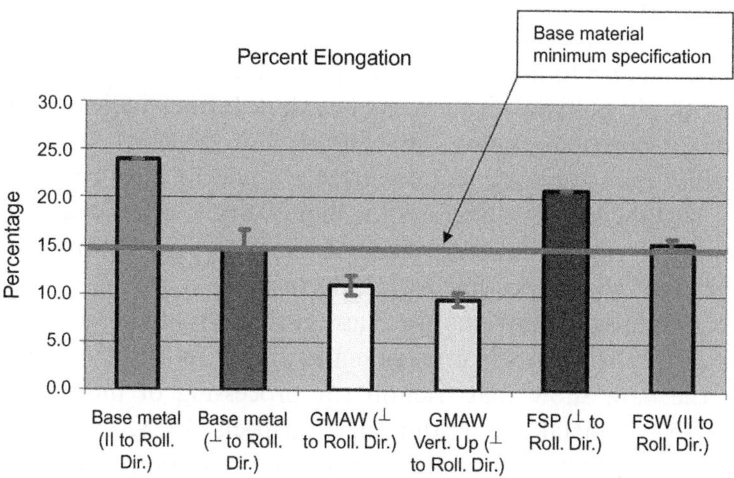


*Figure 4.27 A comparison of elongation for all the sample types tested. Also marked is the minimum specification for the base material.*

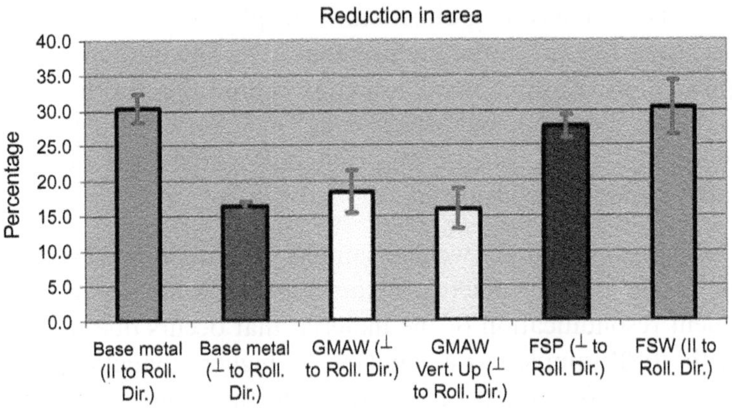


*Figure 4.28 A comparison of reduction of area for all the sample types tested.*

The reduction of area was also calculated. This was determined by measuring the width and thickness of posttensile tested samples with calipers and then multiplying the two values. The resulting data are shown in Figure 4.28. The data show similar trends as the elongation data.

## 4.10.3 Summary

Macro tensile testing data were reviewed for all of the flat, nonformed samples. The data highlight the differences between the base material and the effects of both gas metal arc welding and friction stir processing. The differences in the data between the samples tested parallel and perpendicular to the rolling direction in the base material highlight the anisotropic material properties. The data further indicate why the traditional approach employs splicing of short sections (width direction) rather than use long sections which would in theory not require splicing by GMAW. Specifically, the elongation is significantly higher parallel to the rolling direction. Second, the data demonstrate the ability of friction stir processing to locally increase ductility, though are not representative of the local increase in ductility due to the samples being partially friction stir process material and mostly base material. Lastly, the data show that friction stir processing of the material is superior to gas metal arc welding in all aspects; UTS, yield strength, elongation, and reduction of area. This could have potential impact or consideration for future structural designs and could allow for gage reductions, resulting in a weight reduction.

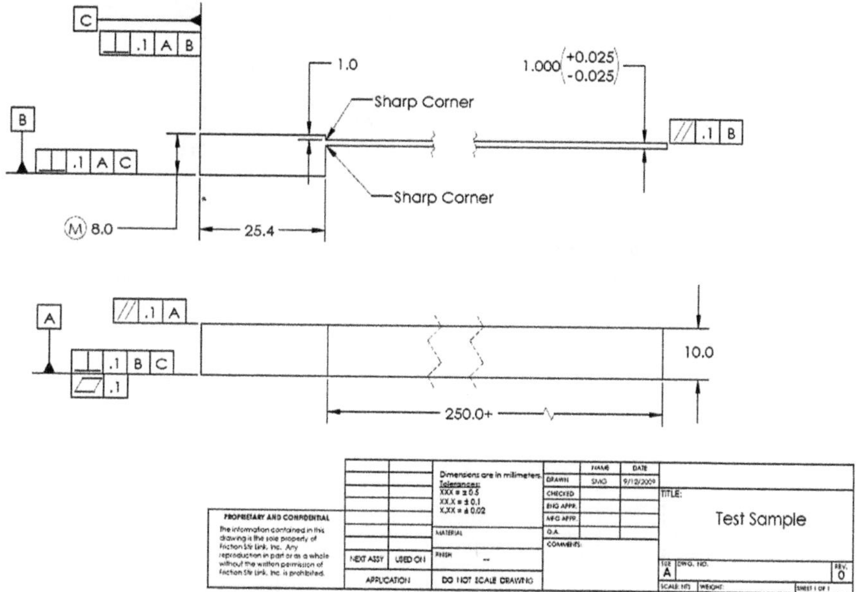


*Figure 4.29 The modified sample shape used in this study.*

## 4.11 YOUNG'S AND SHEAR MODULUS TEST RESULTS

### 4.11.1 Introduction

Young's and shear modulus data were calculated from vibratory testing performed to ASTM E756 (2005). The test requires a thin beam with a solid end, which is vibrated using a range of excitation frequencies during the test. Data acquisition equipment is then used to measure the response of the beam. With these data, the actual average thickness, and actual length, both Young's and shear moduli can be calculated.

The test nominally requires a beam with a solid end with the material to be tested extending as a beam from the center of the solid end. Due to the friction stir processed region being at the top of the material, this optimal sample type was not possible. The modified sample shape that was generated, with agreement of the testing facility, is shown in Figure 4.29. The extended beam is located between 1 and 2 mm below the surface. For consistency, all testing reviewed used sample types of the same shape. Due to the requirements for a specific beam shape, only data from flat samples could be reviewed. The sample types included base metal, friction stir processed, and gas metal arc welded samples.

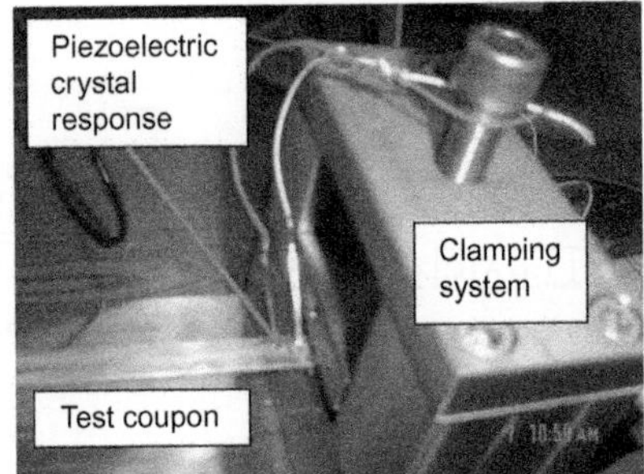


*Figure 4.30 The test apparatus used in this study.*

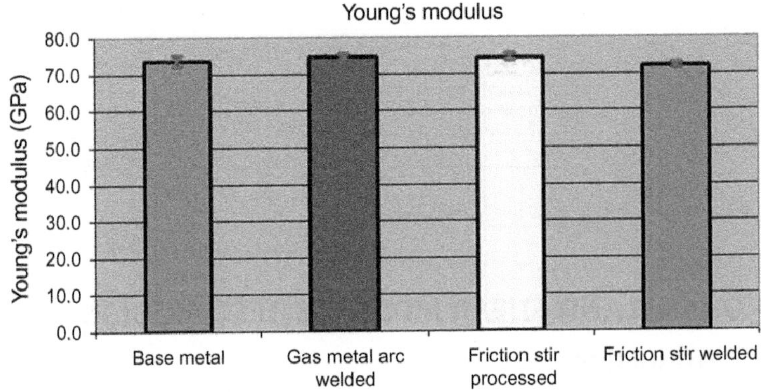


*Figure 4.31 Young's modulus data for the base metal, gas metal arc welded samples, and friction stir processed samples.*

The test apparatus is shown in Figure 4.30. The clamping system is shown, as well as the beam excitation device. Also, the piezoelectric crystal, used to send the vibratory response of the beam, is shown on the right.

### 4.11.2 Test Results

Data from the tests are shown in Figure 4.31. Data for the base metal, gas metal arc welded samples, and friction stir processed samples are shown. There were a total of three samples for each sample type and six excitation frequencies for each sample. This yielded 18 total datum points for each sample type. The data were averaged and standard deviations were calculated. The values ranged from about 73–75 GPa, which matches expected values. In addition, there is no statistically significant difference between the various types, indicating that both FSP and GMAW have no influence on Young's modulus, as would be

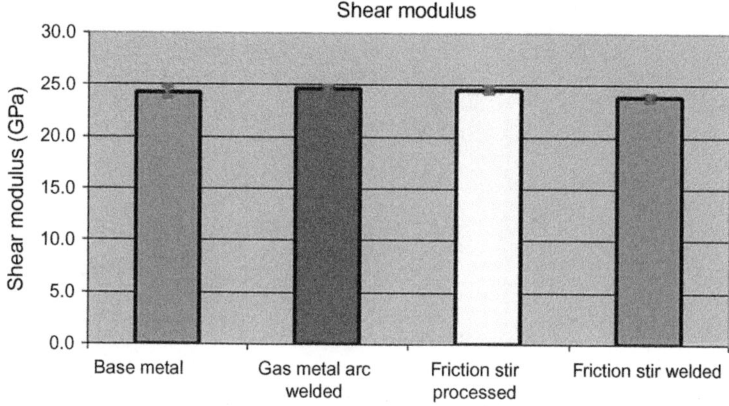


*Figure 4.32 Shear modulus data for the base metal, gas metal arc welded samples, and friction stir processed samples.*

expected. Modulus of metallic alloys is chemistry dependent, but does not depend on grain size. So, defect-free 5XXX alloy in GMAW and FSP microstructural states are expected to have similar modulus values.

Shear modulus data calculated from the same ASTM E756 tests are shown in Figure 4.32. The data show similar trends as the Young's modulus data, given the same test is used to calculate shear modulus. Values were measured to be approximately 24.5 GPa. This is within the range of what would be expected. In addition, there does not appear to be any statistically significant difference between the base material and the friction stir processed or gas metal arc welded samples. Again, this is consistent with theoretical expectations as discussed above.

### 4.11.3 Summary

Young's and shear modulus data were reviewed which were calculated from vibratory measurement data from testing performed per ASTM E756. Data were generated for flat samples only. There was no statistical difference between the base material and either friction stir welded, friction stir processed or gas metal arc welded samples, indicating no process has an effect on Young's or shear modulus.

## 4.12 MINI TRANSVERSE TENSILE TEST RESULTS

### 4.12.1 Introduction

To determine local material properties through the thickness of the base metal as well as in and around the gas metal arc welded and

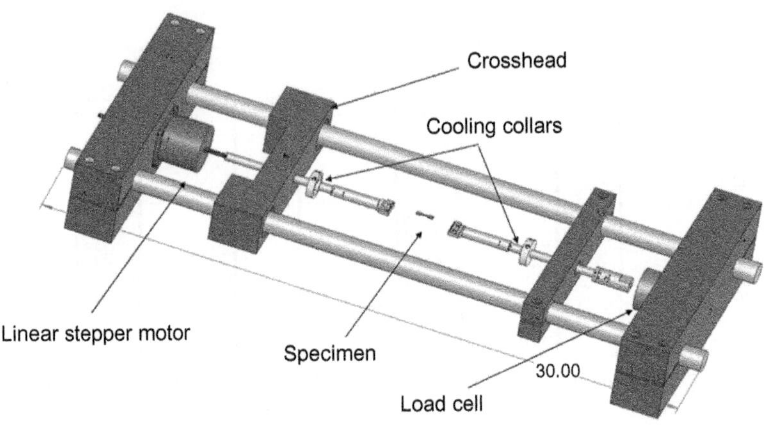


*Figure 4.33 The mini tensile test apparatus used in this study.*

friction stir processed zones, mini tensile testing is required. In this particular chapter, data are reviewed for tensile testing that is transverse to the weld or processing directions. As described in the sample type description, friction stir processing was performed parallel to the rolling direction and the gas metal arc welding was performed perpendicular to the rolling direction. Thus, the transverse tensile tests were performed perpendicular to the rolling direction for the FSP samples and parallel to the rolling direction for the GMAW samples. There are no accepted standards for this testing, but testing was performed at the Missouri University of Science and Technology per their internal procedure (Mishra, 2008). This procedure has been used for numerous material research studies with data reported in many publications.

The test apparatus is shown in Figure 4.33. It consists of equipment that would be on typical tensile testing machines, but on a smaller scale. There is a stepper motor coupled to a crosshead whose resulting linear motion is used to pull the sample in tension. Connected to the crosshead are cooling collars (the setup can be used for high temperature testing) and subsequently one set of grips. On the stationary side, there is a second cooling collar and second set of grips, which are mounted to a load cell. The load cell is connected to a data acquisition system which records load versus strain, as is typical of standard tensile testing machines. The extension is calculated from the crosshead movement based on the programmed displacement of stepper motor.

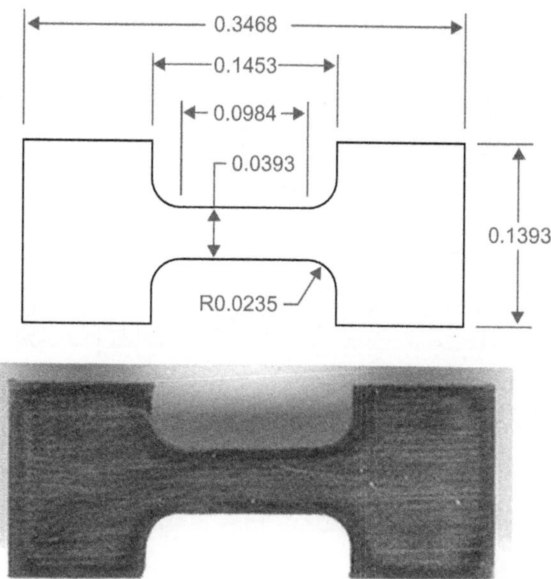


*Figure 4.34 The mini tensile specimen geometry used in this study.*

A standard tensile test coupon has been developed as part of the procedure. The coupon geometry is shown in Figure 4.34. The gage length is selected between 1.3 and 5 mm depending on the study and the specimen thickness is kept between 0.5 and 1 mm. For this case study, the gage length of the samples was 2.5 mm, the thickness was 0.5 mm, and the width was 1 mm. These coupons were carefully removed from the subject samples and machined with a mini CNC milling machine. These samples were first machined with a 1 mm diameter end mill with thickness significantly greater than the gage thickness. The thick sample, machined to the outline of the test coupon, was then carefully cut using a low speed diamond saw. Finally, the sample was polished on lapping films to the final desired gage thickness. The location of the test coupons from the test samples is tightly controlled.

The elongation is measured from crosshead displacement data from the programmed stepper motor movement. The machine is calibrated with physical measurement of deformed specimens and LVDT measurement of crosshead movement. As a cross-check, a few specimens were photographed and calculated elongation compared with physical measurements.

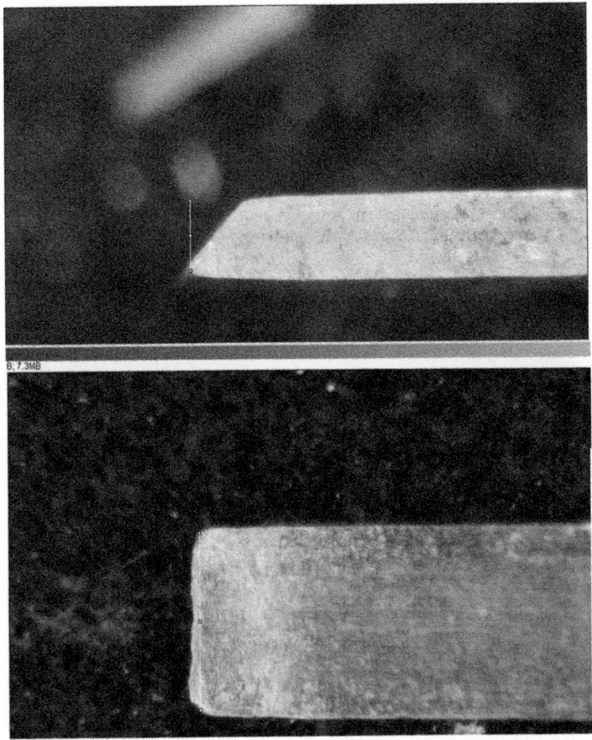

*Figure 4.35 Images of the fracture surface (originally photographed at 100× magnification) from the top and bottom. Such images were used for calculation of reduction of area.*

With mini tensile testing, measurements for calculating the reduction of area are necessarily different than that with macro tensile testing. The reduction in length and width measurements used for calculating the reduction of area are small enough that calipers are not a practical tool for measurement. A special procedure has been developed for this purpose. The first step is to capture an image of the fracture surface at 100× magnification from the top and bottom (see Figure 4.35). Special imaging software with the ability to trace the length and width at the necked portion of the tensile tested sample is used. With the aid of a micron calibration bar, the system estimates the thickness and width of the sample. These values are then used for the reduction of area calculations.

### 4.12.2 Test Results

Transverse mini tensile testing was performed for all sample types. However, the locations and number of mini tensile tests varied

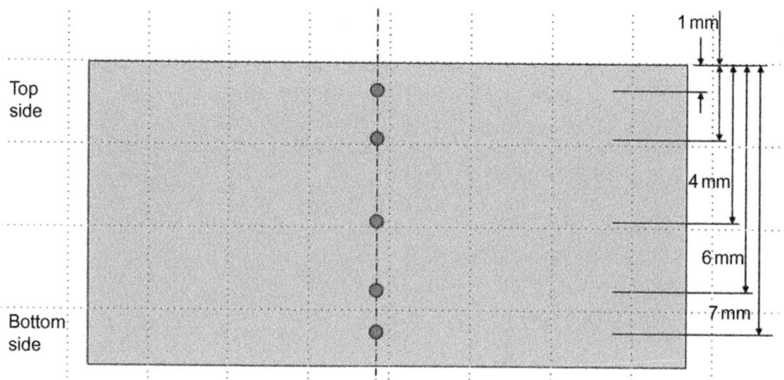


*Figure 4.36 The locations of the mini tensile tests for the unformed base material (sample type A).*

depending on the type of sample. For each sample type and location, three mini tensile tests were performed to understand the relative level of repeatability. The mini tensile testing was performed per the procedure described in the previous section. UTS, yield strength, elongation, and reduction of area were either measured or calculated from each of the mini tensile tests.

### 4.12.2.1 Unformed Base Metal

The locations of the mini tensile tests for the unformed base material (sample type A) are shown in Figure 4.36. It can be seen that there were five locations through the thickness where mini tensile tests of the unformed base material were performed. Each of the three samples for the base material was tested in these locations, for a total of 15 mini tensile tests.

Figure 4.37 displays the UTS results from the mini tensile tests of the base material. The data for each location is placed next to that location. Results from all three tests are included as well as the average and standard deviation. It can be seen that the UTS data is relatively consistent through the thickness of the material. It is also noted that the average values are somewhat less and the standard deviation (variation) is somewhat greater than that indicated by the macro tensile tests.

These trends are generally expected due to the relative size of the coupons and the relative effect of stress risers (e.g., corners) and any other imperfections. The volume of material averaged during testing is obviously smaller in a mini specimen. Therefore, any impact of microstructural inhomogeneities, such as iron containing constituent

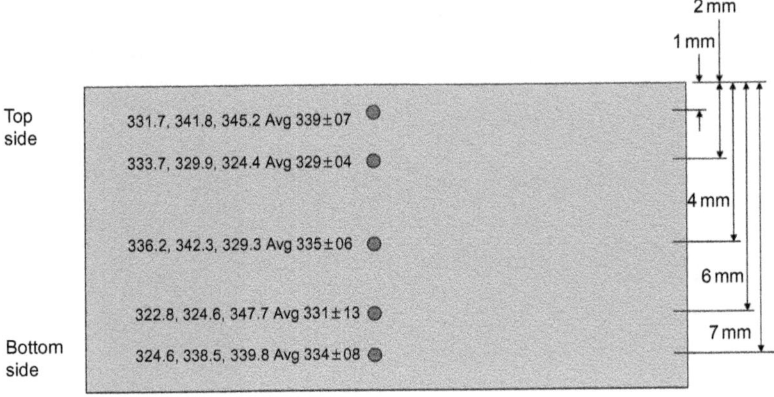


*Figure 4.37 UTS values from the mini tensile tests of the base material.*

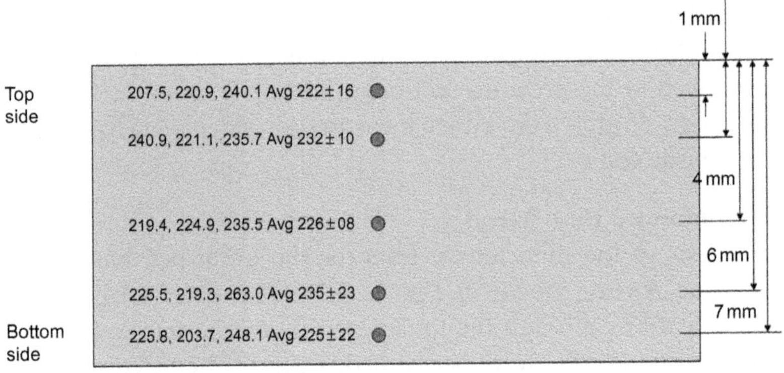


*Figure 4.38 Yield strength values of the base material from the mini tensile tests.*

particles, becomes magnified. Since the main objective of this study is to compare the results of the various sample types, the general offset (lower values) are not an issue. However, the increased variation can have some impact, since it will slightly lower the statistical confidence in conclusions regarding any differences between the various sample types.

Figure 4.38 displays the yield strength (YS) results from the mini tensile tests of the base material. It can be seen that the YS data is relatively consistent through the thickness of the material. Similar to the UTS data, it is observed that the average values are somewhat less and the standard deviation (variation) is somewhat greater than that indicated by the macro tensile tests. This will be for the same reasons as noted above.

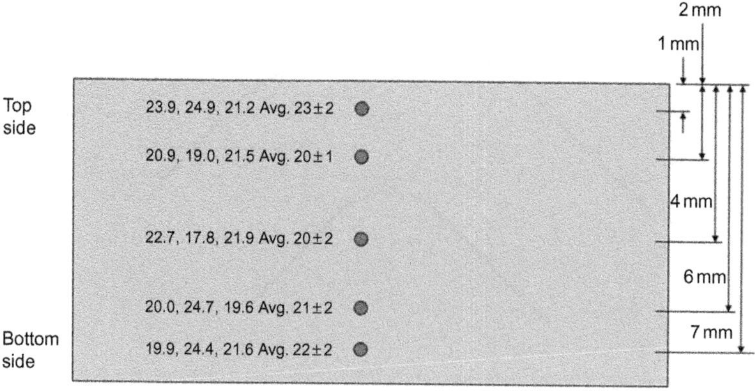


*Figure 4.39 Elongation values from the mini tensile tests of the base material.*

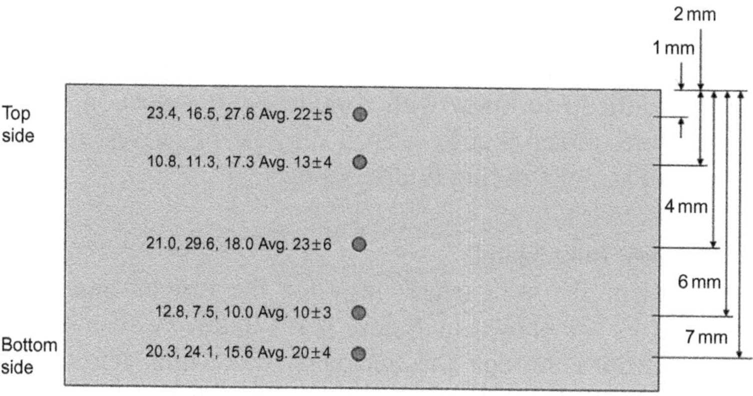


*Figure 4.40 Reduction of area results from the mini tensile tests of the base material.*

Figure 4.39 displays the elongation results from the mini tensile tests of the base material. It can be seen that the elongation data is relatively consistent through the thickness of the material. Similar to the UTS and YS data, it is observed that the average values are somewhat less and the standard deviation (variation) is somewhat greater than that indicated by the macro tensile tests. The cause of this will be for the same reasons as previously discussed.

Figure 4.40 displays the reduction of area results from the mini tensile tests of the base material. It would appear that the reduction of area data is more variable through the thickness of the material. However, there is more variation in the data. Thus, it is more difficult to make conclusions based on any trends seen in the data. It is seen

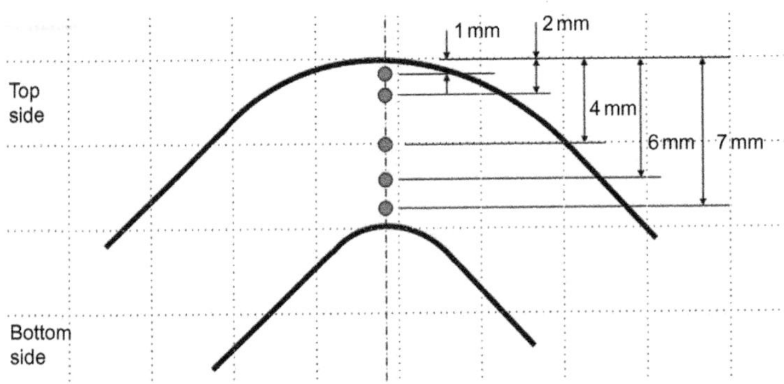


*Figure 4.41 The locations of the mini tensile tests for the formed base material (sample type F).*

that the overall average is fairly similar to the base material. The variation or standard deviation is likely greater due to the more difficult and likely more variable measurement technique for the area of the fractured surface. This may also impact the average value as well, though it is difficult to know with confidence. If reduction of area is deemed a more critical metric, then it may be of interest to increase the number of samples during future testing.

#### 4.12.2.2 Formed Base Metal

The locations of the mini tensile tests for the formed base material (sample type F) are shown in Figure 4.41. It can be seen that there were five locations through the thickness where mini tensile tests of the formed base material were performed. These locations were in the center of the formed region (1.25″ inner diameter) and matched the locations of the unformed base material. Each of the three samples for the base material was tested in these locations, for a total of 15 mini tensile tests.

Figure 4.42 displays the UTS results from the mini tensile tests of the formed base material. Similar to the previous figures, the data for each location is placed next to that location. Results from all three tests are included as well as the average and standard deviation. It can be seen that the UTS data is not as consistent through the thickness of the material as compared with the base material. It is also noted that the average values are somewhat greater than the unformed base material, as would be expected. This is due to the work hardening effect of the forming operation. The variation through the thickness is likely a result of the varied amount of work hardening, with greatest being on

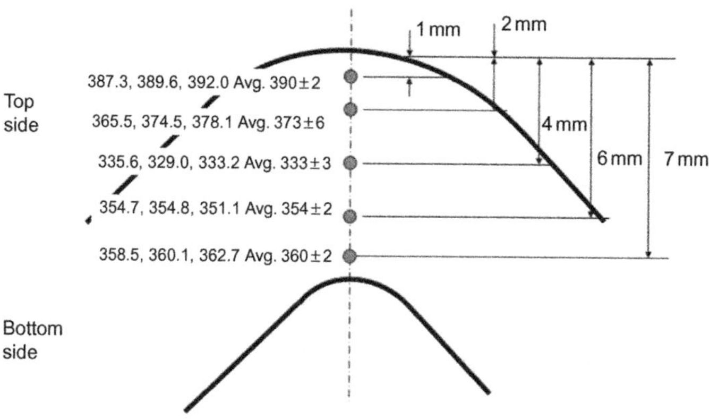


*Figure 4.42 UTS values from the mini tensile tests of the formed base material.*

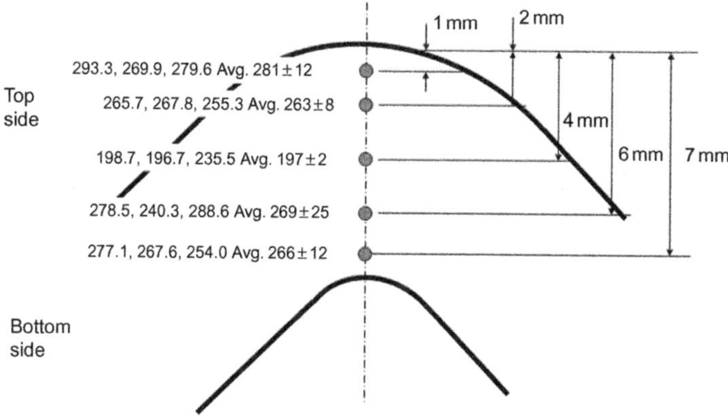


*Figure 4.43 Yield strength values from the mini tensile tests of the formed base material.*

the outer radius. The forming strain varies from the surface to center, being maximum at the surface. It is also seen that the variation or standard deviation is fairly similar to that of unformed base material, as would be expected.

Figure 4.43 displays the yield strength (YS) results from the mini tensile tests of the formed base material. It is seen that the YS data shows greater variation through the thickness of the material, with the minimum being in the center. Similar to the UTS data, it is observed that the average values are somewhat higher than that of the unformed base metal. Again, this is expected and is due to the work hardening effect of the forming operation. In addition, the standard deviation (variation) is similar to the unformed base metal, raising no concerns.

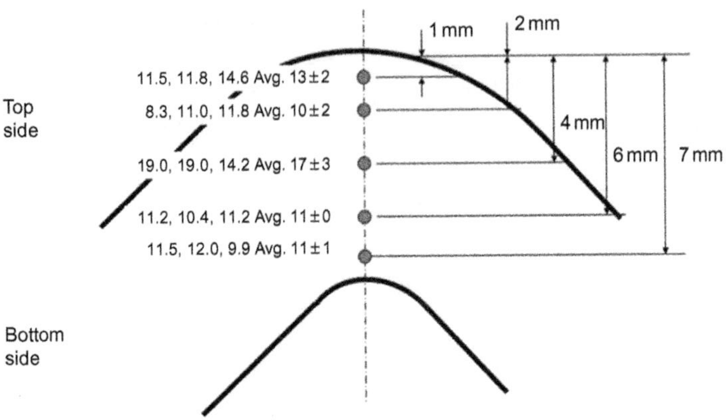


*Figure 4.44 Elongation values from the mini tensile tests of the base material.*

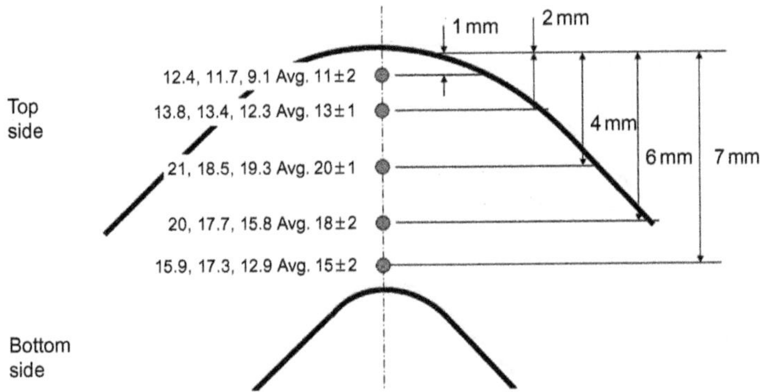


*Figure 4.45 Reduction of area values calculated from the mini tensile tests of the formed base material.*

Figure 4.44 displays the elongation results from the mini tensile tests of the base material. It is seen that the elongation data have similar variation through the thickness of the material as the UTS and YS data. It is observed that the average values are somewhat less than the unformed base metal. As with the UTS and YS, this is a result of the work hardening effect of the forming operation. Lastly, the standard deviation (variation) is similar to that of the unformed base metal.

Figure 4.45 displays the reduction of area results from the mini tensile tests of the formed base material. As with the base material, the reduction of area data is more variable through the thickness of the material. It is seen that the average is lower than the base material. This is expected and is for the same reason as the differences in the other material property characteristics.

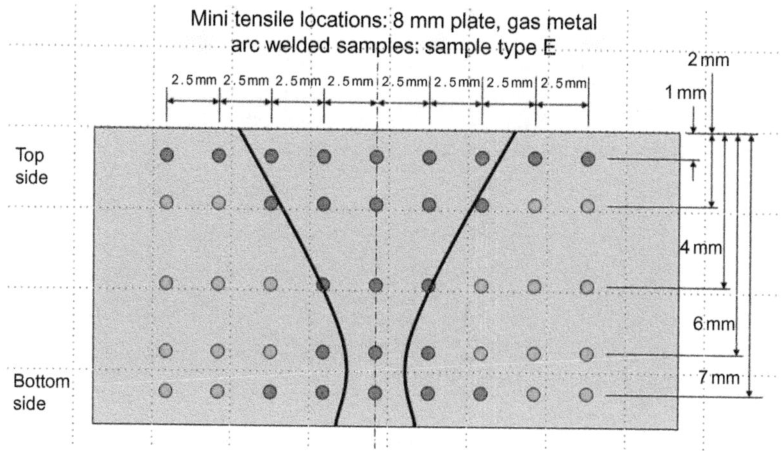


*Figure 4.46 The locations of the mini tensile tests for the gas metal arc weld samples (sample type E).*

### 4.12.2.3 Gas Metal Arc Weld

The locations of the mini tensile tests for the gas metal arc weld samples (sample type E) are shown in Figure 4.46. It can be seen that there were five locations through the thickness and nine locations transverse to the welding direction for a total of 45 measurement locations where mini tensile tests of the gas metal arc weld samples were performed. The locations transverse to the weld direction were selected at distance such that samples would extend into the base material, so as to generate data in the HAZ, as well. In the locations indicated by red dots, mini tensile tests were performed for all three samples, whereas in the locations indicated by the green dots, mini tensile tests were only performed for one of the samples.

Figure 4.47 displays the UTS results from the mini tensile tests of the gas metal arc welded samples. Similar to the previous figures, the average and standard deviation data for each location is placed next to that location. However, the individual data points are not shown to avoid cluttering the diagram. It can be seen that the average UTS data has much more variation as a function of location than the base material. The values range from near the base material values (325 MPa) to as low as approximately 260 MPa. These lowest values are somewhat less than the data generated from the macro tensile tests. As discussed, the relative size of the coupon and the impacts of the corners and any imperfections reasonably explain this deviation. It is also seen that the variation or standard deviation is fairly similar to or slightly greater

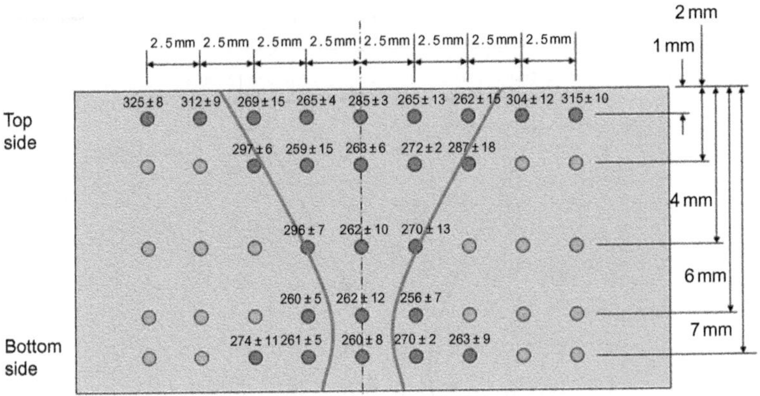


*Figure 4.47 UTS values from the mini tensile tests of the gas metal arc welded samples.*

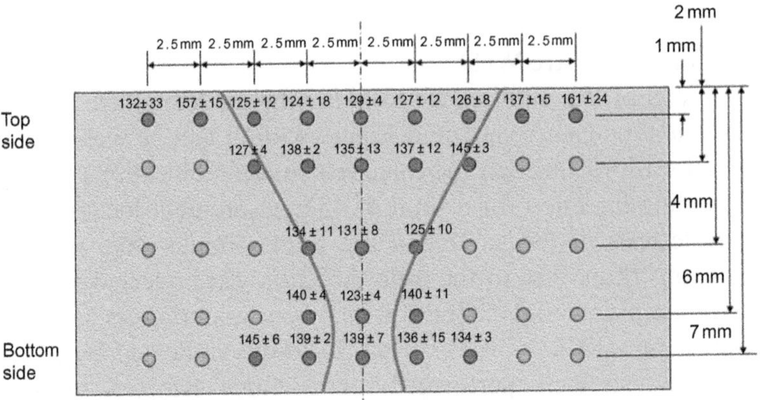


*Figure 4.48 Yield strength values from the mini tensile tests of the gas metal arc welded samples.*

than that of unformed base material. It is not surprising that there would be a greater variation, especially in the weld nugget, since the effect of any porosity would have a much greater impact on the mini tensile tests.

Figure 4.48 displays the yield strength (YS) results from the mini tensile tests of the GMAW samples. It can be seen that the YS data has similar trends versus location as the UTS data. The highest values are farthest from the weld nugget and are lowest in and around the weld nugget. However, the highest values do not approach the base material properties nearly as closely as the UTS data. This indicates that the HAZ, from a YS perspective, is larger than for the UTS.

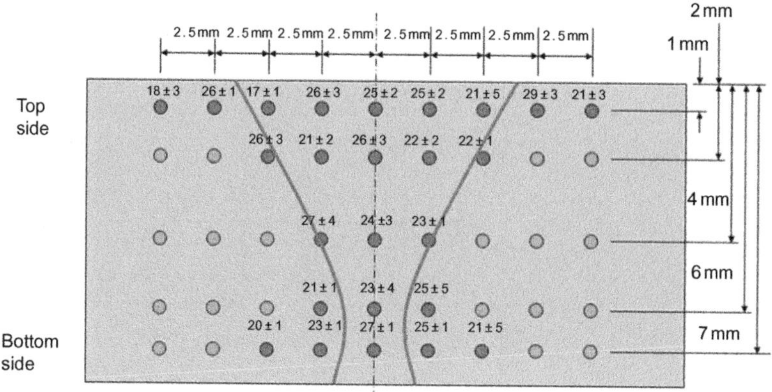


*Figure 4.49 Elongation values from the mini tensile tests of the gas metal arc welded samples.*

The variation or standard deviation is similar to the UTS data in the weld nugget, but larger in the HAZ. This indicates the weld width or heat input may be changing versus location of weld sample. This is indicative of the variability of the manual welding process, and is not unexpected. The data also show a consistent trend as the macro tensile data. The measured values for this particular metric are lower than from the macro tensile samples.

Figure 4.49 displays the elongation results from the mini tensile tests of the GMAW samples. It can be seen that the elongation data is relatively consistent throughout the area in and around the weld nugget, with the lower of the values near the unaffected base metal and near the surface. It is observed that the average values are similar to the base metal in the direction perpendicular to the rolling direction, which is the same testing direction as these tests. This indicates the GMAW has limited local effect on the elongation. Lastly, the standard deviation (variation) is similar to that of the unformed base metal, which indicates a stable and consistent process.

Figure 4.50 displays the reduction of area results from the mini tensile tests of the GMAW samples. The reduction of area data has some variation in and around the weld nugget, with the lowest values near the surface. It is seen that the average is about the same as the base material.

#### 4.12.2.4 Unformed Friction Stir Processed Samples

The locations of the mini tensile tests for the unformed friction stir processed samples (sample type B) are shown in Figure 4.51. It can be

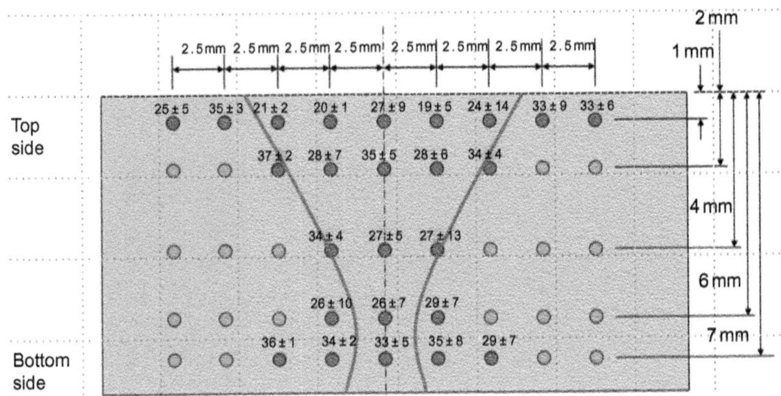


*Figure 4.50 Reduction of area values from the mini tensile tests of the gas metal arc welded samples.*

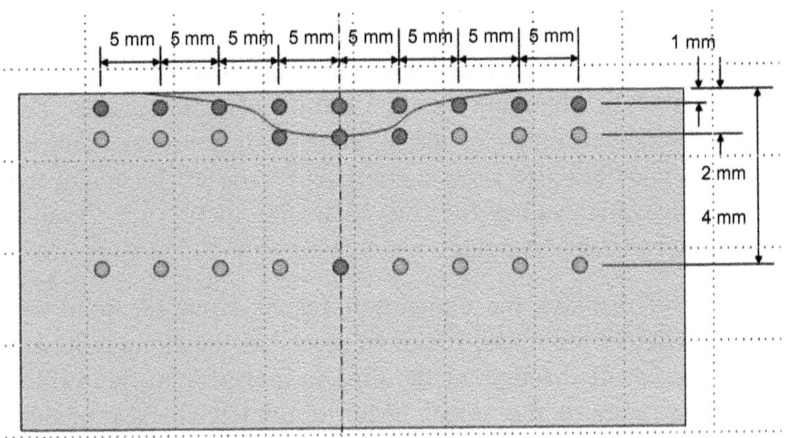


*Figure 4.51 The locations of the mini tensile tests for the unformed friction stir processed samples (sample type B).*

seen that there were three locations through the thickness and nine locations transverse to the friction stir processing direction, for a total of twenty-seven locations where mini tensile tests of the unformed friction stir processed samples were performed. Only three vertical locations were selected due to the shallower depth of the FSP versus the full depth of the GMAW. Similar to the GMAW samples, the locations transverse to the processed direction were selected at distance such that samples would extend into the base material, so as to generate data in the HAZ, as well. In the locations indicated by red dots, mini tensile tests were performed for all three samples whereas the

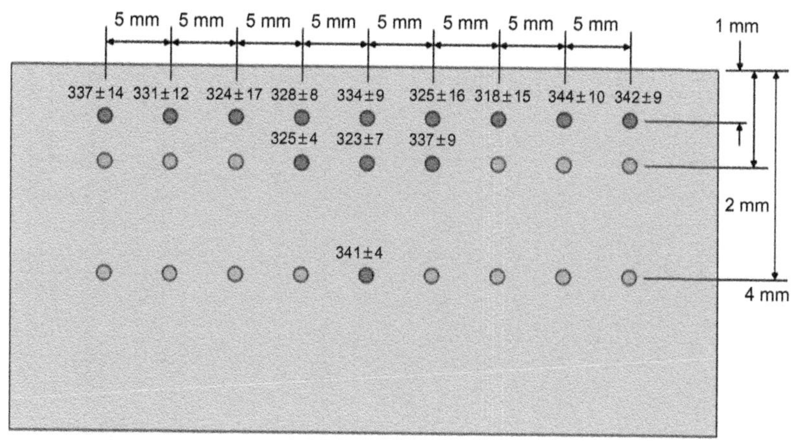


*Figure 4.52 UTS values from the mini tensile tests of the unformed friction stir processed samples.*

locations indicated by the green dots, mini tensile tests were only performed for one of the samples.

Figure 4.52 displays the UTS results from the mini tensile tests of the unformed friction stir processed samples. Similar to the GMAW charts, the average and standard deviation data for each location is placed next to that location. It can be seen that the average UTS data has much less variation as a function of location versus the gas metal arc welded samples and similar to that of the base material. The values range from near the base material values (345 MPa) to as low as approximately 323 MPa, compared to as low as 260 MPa for the GMAW samples. These values are somewhat less than the data generated from the macro tensile tests. As discussed, the relative size of the coupon and the impacts of the corners and any imperfections reasonably explain this deviation. It is also seen that the variation or standard deviation is fairly similar to that of unformed base material and lower than the GMAW samples. This indicates the relative consistency and benefits of friction stir processing.

Figure 4.53 displays the yield strength (YS) results from the mini tensile tests of the unformed FSP samples. It is seen that the YS data shows larger variation than the UTS data, which is a similar trend as the GMAW samples. This indicates that the HAZ, from a YS perspective, is larger than for the UTS. The highest values are farthest from the stir zone and are lowest in and around the stir zone. However, unlike the GMAW samples, the highest values approach the base

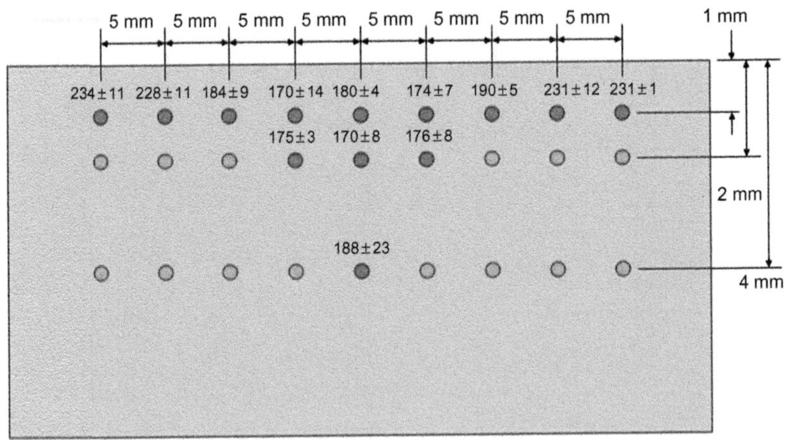


*Figure 4.53 Yield strength values from the mini tensile tests of the unformed friction stir processed samples.*

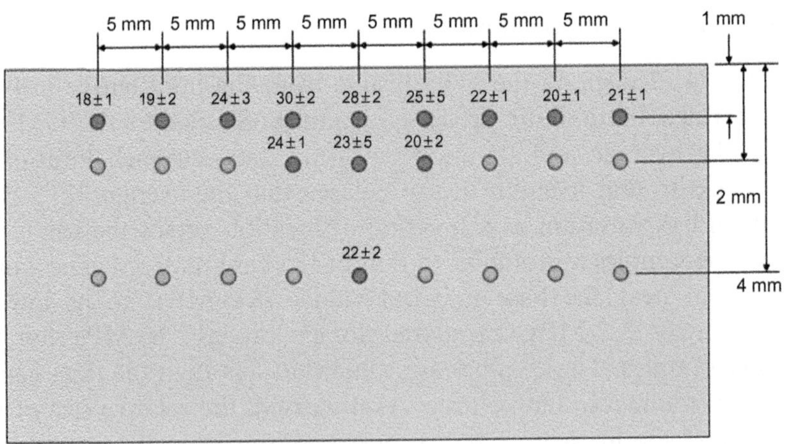


*Figure 4.54 Elongation values from the mini tensile tests of the unformed friction stir processed samples.*

material properties, indicating the benefit of the lower heat input of FSP. The standard deviation is similar to the UTS data. The data also show a consistent trend, but greater difference, with respect to the macro tensile samples. The relative difference for this particular metric is greatest in the stir zone, likely due to the more local nature of FSP (only 2 mm depth).

Figure 4.54 displays the elongation results from the mini tensile tests of the unformed FSP samples. It can be seen that the elongation

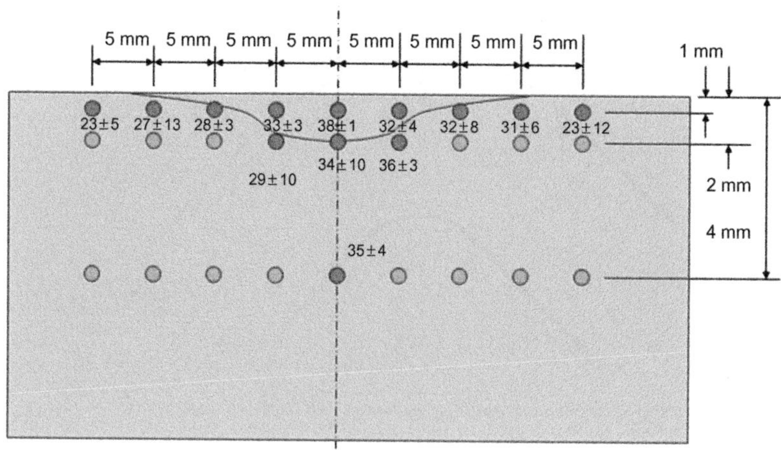


*Figure 4.55 Reduction of area values from the mini tensile tests of the unformed friction stir processed samples.*

data is highest in the stir zone, with the lower of the values tending to be closest to the unaffected base metal and near the surface. As expected, away from the stir zone, the average values are similar to the base metal in the direction parallel to the rolling direction, which is the same testing direction as these tests. However, there are significant increases in the elongation in the stir zone. This indicates that FSP has a significant local and positive effect on the elongation. Lastly, the standard deviation (variation) is similar to that of the unformed base metal, which indicates a stable and consistent process.

Figure 4.55 displays the reduction of area results from the mini tensile tests of the unformed FSP samples. The reduction of area data shows similar trends in and around the stir zone as with the elongation data. The lowest values are away from the stir zone.

#### 4.12.2.5 Formed FSP Samples

The locations of the mini tensile tests for the formed FSP samples (sample type D) are shown in Figure 4.56. These samples were formed to the 5/16″ (8 mm) internal radius. It can be seen that there were three locations through the thickness and nine locations transverse to the processing direction, for a total of twenty-seven locations where mini tensile tests of the formed FSP samples were performed. As with the unformed samples, only three vertical locations were selected due to the shallower depth of the FSP versus the full depth of the GMAW. Otherwise, the locations transverse to the processed direction were in

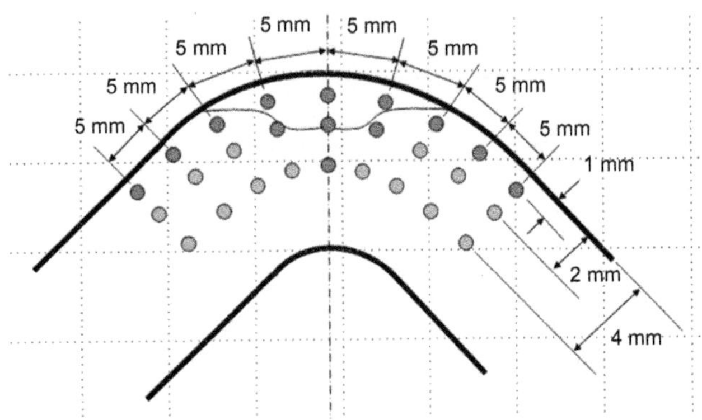


*Figure 4.56 The locations of the mini tensile tests for the formed FSP samples (sample type D).*

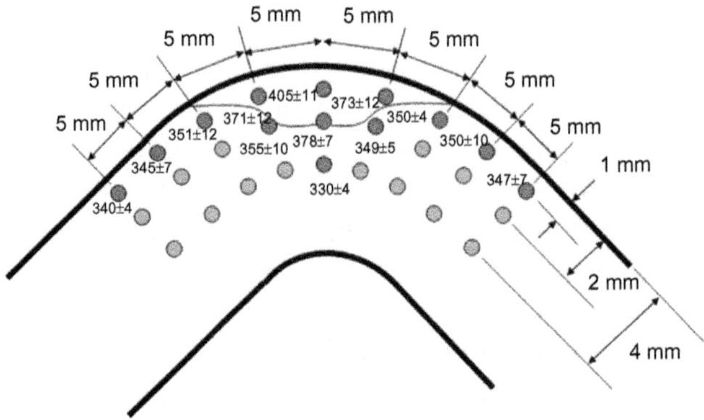


*Figure 4.57 UTS values from the mini tensile tests of the formed friction stir processed samples.*

the same relative locations as with the unformed material, but in the formed state. In the locations indicated by red dots, mini tensile tests were performed for all three samples whereas the locations indicated by the green dots, mini tensile tests were only performed for one of the samples.

Figure 4.57 displays the UTS results from the mini tensile tests of the formed friction stir processed samples. Similar to the previous figures, the average and standard deviation data for each location is placed next to that location. Compared with the unformed FSP samples, it can be seen that the UTS values increase. In fact, they meet or exceed the base material properties. The values range from 405 MPa near the surface to a low of approximately 330 MPa near the

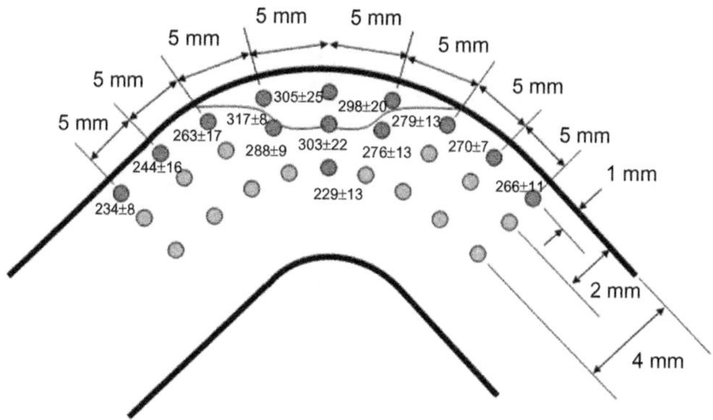


*Figure 4.58 Yield strength values from the mini tensile tests of the formed friction stir processed samples.*

center, compared to as low as 260 MPa for the GMAW samples and approximately 335 MPa for the base material. The relative increase in UTS is caused by the work hardening effect of the forming operation. It is also seen that the variation or standard deviation is fairly similar to that of unformed base material and lower than the gas metal arc welded samples. This indicates the relative consistency and benefits of friction stir processing.

Figure 4.58 displays the yield strength (YS) results from the mini tensile tests of the formed FSP samples. Compared to the unformed FSP samples, it can be seen that the YS values increase, and increase more than the UTS. Furthermore, they meet or exceed the base material properties. The values range from 305 MPa near the surface to a low of approximately 230 MPa near the center, compared to as low as 125 MPa for the GMAW samples and approximately 230 MPa for the base material. As before, the relative increase in YS is caused by the work hardening effect of the forming operation. It is also seen that the variation or standard deviation is fairly similar to that of unformed base material and lower than the GMAW samples. This indicates the relative consistency and benefits of FSP.

Figure 4.59 displays the elongation results from the mini tensile tests of the formed FSP samples. As compared to the unformed FSP samples, the values are lower. This is a result of the work hardening effect of the forming process. It can be seen that the elongation data is lowest near the surface where the work hardening effect is greatest. The higher values are near the center and away from the formed area.

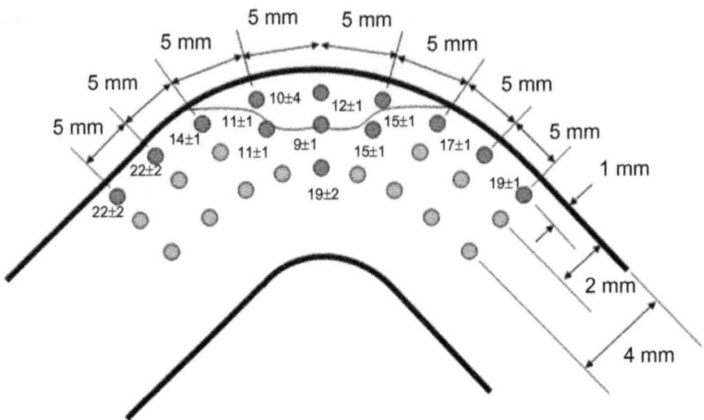


*Figure 4.59 Elongation values from the mini tensile tests of the formed friction stir processed samples.*

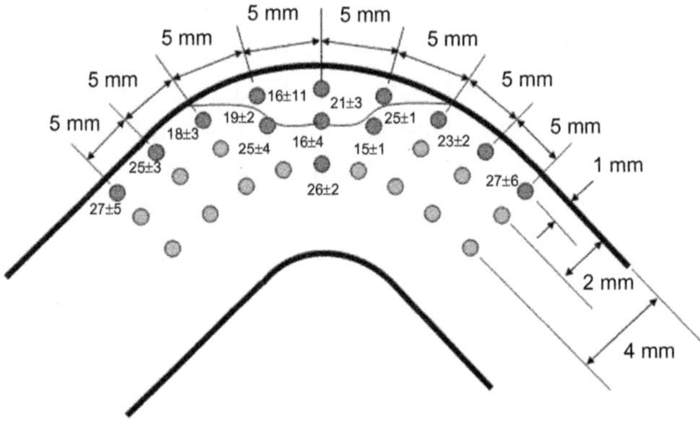


*Figure 4.60 Reduction of area values from the mini tensile tests of the formed friction stir processed samples.*

Lastly, the standard deviation (variation) is similar to that of the unformed base metal, which indicates the stability and benefits of FSP.

Figure 4.60 displays the reduction of area results from the mini tensile tests of the formed FSP samples. As with most of the other data, they show similar trends to the elongation data.

#### 4.12.2.6 Friction Stir Welded Samples

The locations of the mini tensile tests for the friction stir welded samples (sample type C) are shown in Figure 4.61. It can be seen that there were five locations through the thickness and nine locations transverse to the processing direction, for a total of forty-five locations where mini tensile tests of the friction stir welded samples were performed.

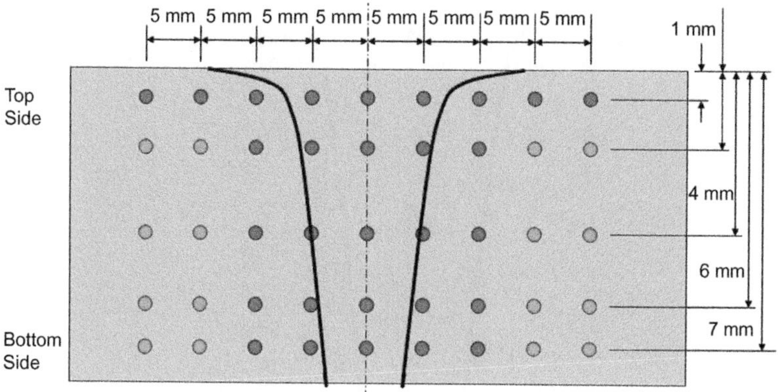


*Figure 4.61 The locations of the mini tensile tests for the friction stir welded samples (sample type C).*

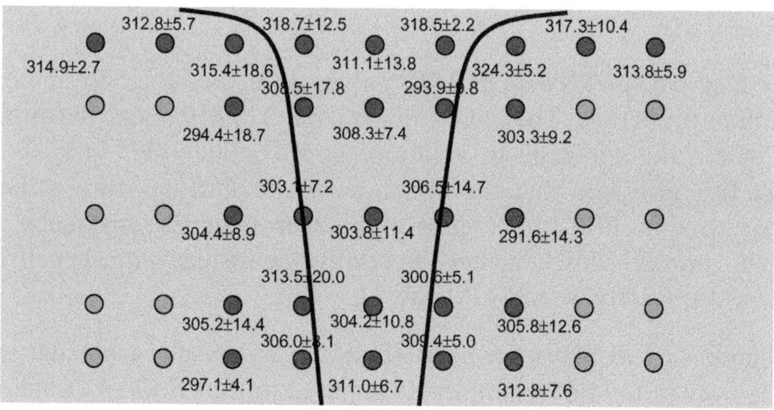


*Figure 4.62 UTS values from the mini tensile tests of the friction stir welded samples.*

Three samples were tested. In the locations indicated by red dots, mini tensile tests were performed for all three samples whereas the locations indicated by the green dots, mini tensile tests were only performed for one of the samples.

Figure 4.62 displays the UTS results from the mini tensile tests of the friction stir welded samples. Similar to the previous figures, the average and standard deviation data for each location is placed next to that location. Compared with the base material, it can be seen that the UTS values decrease, while compared to GMAW, the results are superior. The values are also somewhat lower than the unformed friction stir processed samples, presumably due to the additional heat

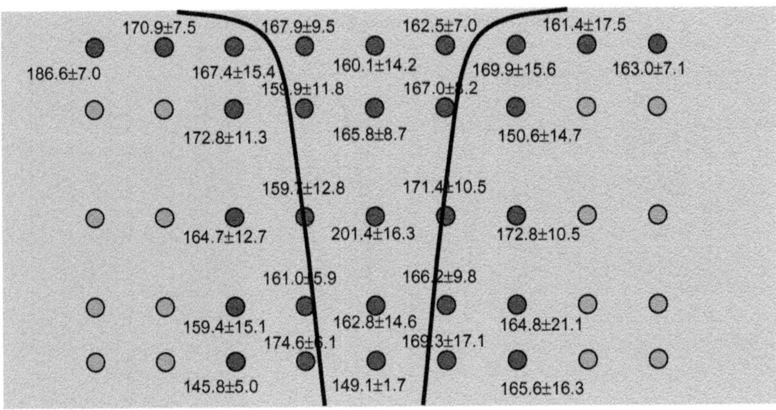


*Figure 4.63 Yield strength values from the mini tensile tests of the friction stir welded samples.*

input required for FSW. The UTS values do exceed the base material minimum properties. The values range from 320 MPa near the surface and within the stir zone to a low of approximately 295 MPa in the HAZ. It is also seen that the variation or standard deviation is fairly similar to that FSP'ed material and lower than the gas metal arc welded samples. This indicates the relative consistency and benefits of friction stir welding versus GMAW.

Figure 4.63 displays the yield strength (YS) results from the mini tensile tests of the FSW samples. Compared to the GMAW samples, it can be seen that the YS values are higher. The values range from near 160 MPa in the HAZ to approximately 200 MPa near the center, compared to as low as 125 MPa for the GMAW samples and approximately 230 MPa for the base material. It is also seen that the variation or standard deviation is fairly similar to that of unformed base material and lower than the GMAW samples. This indicates the relative consistency and benefits of FSW.

Figure 4.64 displays the elongation results from the mini tensile tests of the FSW samples. The FSW samples show a similar trend to the FSP samples. It can be seen that the elongation data within the stir zone is lowest near the center while higher values exist near the top and bottom. It is also seen that the elongation within the stir zone is also higher than the base material. Lastly, the standard deviation (variation) is similar to that of the FSP samples.

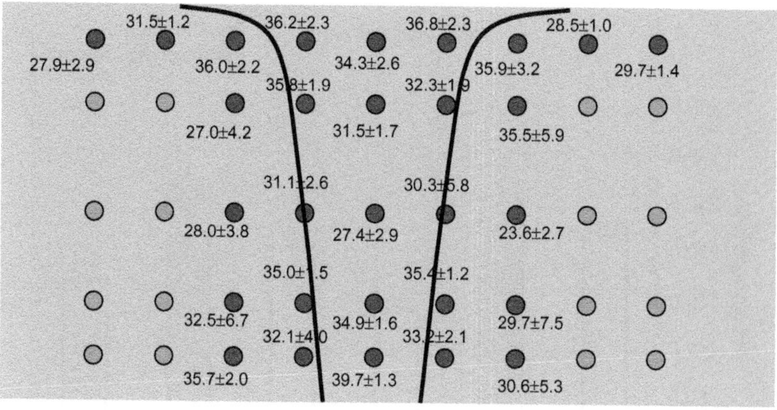


*Figure 4.64 Elongation values from the mini tensile tests of the friction stir welded samples.*

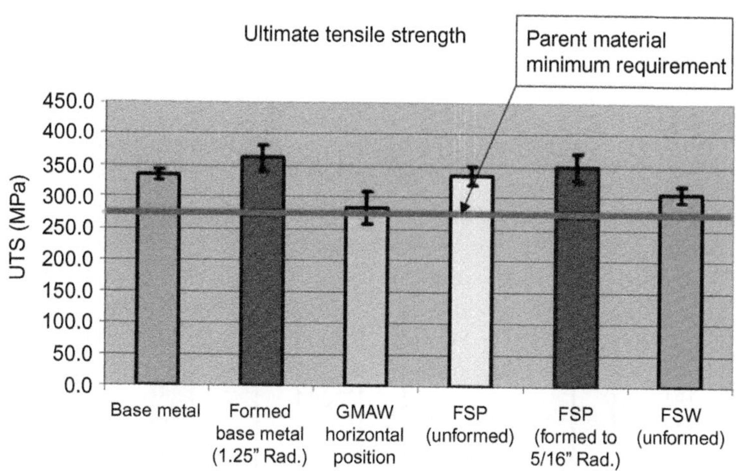


*Figure 4.65 A comparison of UTS in various conditions.*

## 4.12.3 Summary

To summarize all of the data, averages and standard deviations for all samples and all locations for each test sample type have been calculated. The ultimate strength data, yield strength data, elongation data, and reduction of area data are plotted in Figures 4.65–4.68, respectively. These summary charts indicate the same trends as the macro tensile tests. The following specific inferences can be made:

1. The GMAW samples are inferior to FSP and FSW samples in all areas. The most dramatic difference is seen in the yield strength

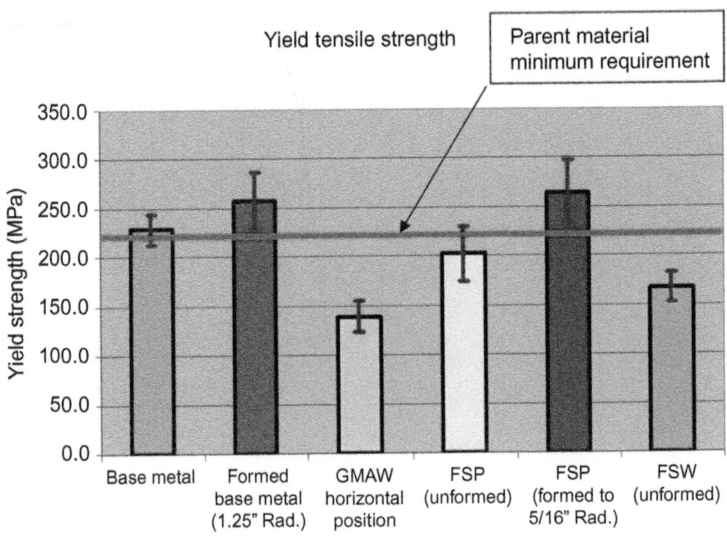


*Figure 4.66 A comparison of yield strength in various conditions.*

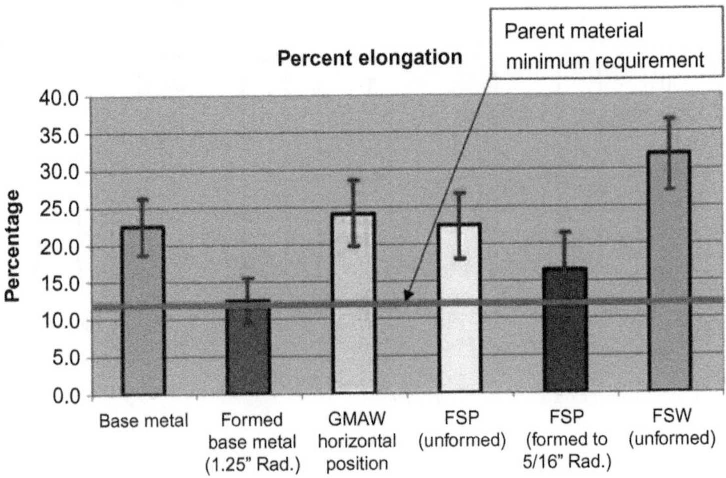


*Figure 4.67 A comparison of elongation in various conditions.*

data. This is important, as structural applications are typically designed to the minimum yield strength.

2. FSP (without forming) has a much smaller effect on the material properties than GMAW and less of an effect than FSW. It can be seen that the UTS is affected very little, approaching parent

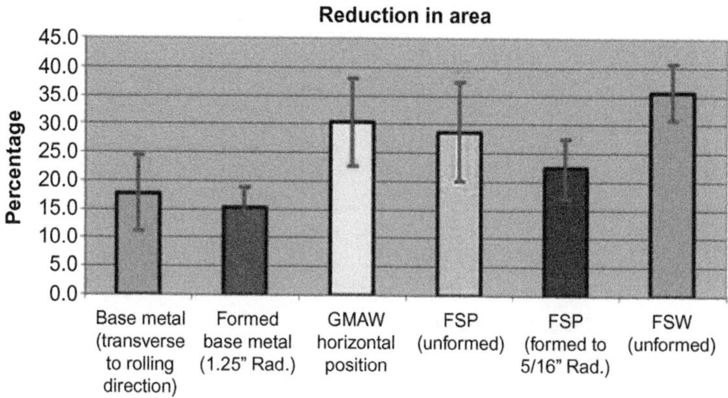


*Figure 4.68 A comparison of reduction of area during mini tensile tests in various conditions.*

material properties. Similar to GMAW, the yield strength is affected most (greatest reduction).

3. FSP has the benefit of locally increasing and improving the ductility, explaining why friction stir processing enables forming to tighter radii. A similar observation is made for FSW, as would be expected, given the similarity between the processes.
4. The forming operation causes work hardening and thus increases in UTS and YS. The elongation and reduction of area are reduced after the forming operations. However, after forming, FSP and subsequently formed samples meet or exceed all parent material minimum properties.

## 4.13 MINI LONGITUDINAL TENSILE TEST RESULTS

### 4.13.1 Introduction

Longitudinal mini tensile testing is reviewed for all sample types. For the review, there were a minimum of two locations where mini tensile testing was performed. The selected locations always included the two lowest or minimum ultimate strength values as calculated from the transverse mini tensile tests. As described in the sample type description, friction stir processing was performed parallel to the rolling direction and the gas metal arc welding was performed perpendicular to the rolling direction. Thus, the longitudinal tensile tests were performed parallel to the rolling direction for the FSP samples and perpendicular to the rolling direction for the GMAW samples. For each sample type and location, three mini tensile tests were performed to understand the

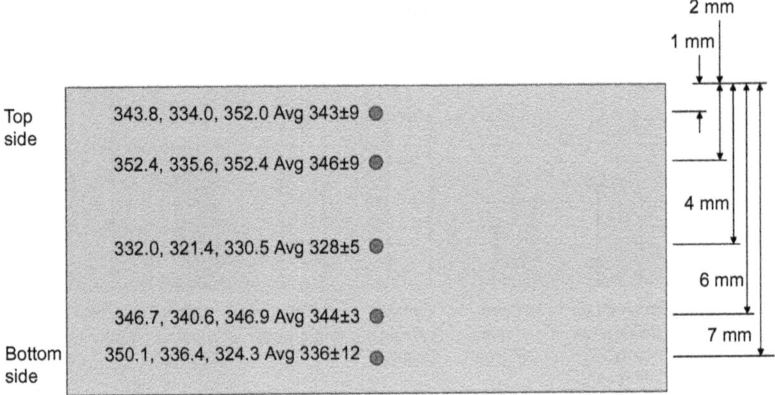


*Figure 4.69 Ultimate strength data in the longitudinal direction (parallel to the rolling direction) for the unformed base material.*

relative level of repeatability. The mini tensile testing was performed per the procedure described in Section 4.5. UTS, yield strength, elongation, and reduction of area were either measured or calculated from each of the mini tensile tests.

### 4.13.2 Results

The ultimate strength data in the longitudinal direction (parallel to the rolling direction) for the unformed base material is shown in Figure 4.69. All five locations tested for the transverse testing were also tested in the longitudinal direction. It can be seen that there are no significant differences in the UTS based on testing direction, by comparing the data below to the data shown in Section 4.5.

The yield strength data in the longitudinal direction for the unformed base material is shown in Figure 4.70. By comparing the data with that shown in Section 4.6, it can be seen that there are no significant differences in the yield strength based on testing direction.

The elongation data in the longitudinal direction (parallel to rolling direction) for the unformed base material is shown in Figure 4.71. By comparing the data with that shown in Section 4.5, it is seen that there is about 3–5% difference in elongation versus the transverse testing direction. This is expected and is somewhat smaller than the difference as measured in the macro tensile testing.

*Figure 4.70 Yield strength data in the longitudinal direction for the unformed base material.*

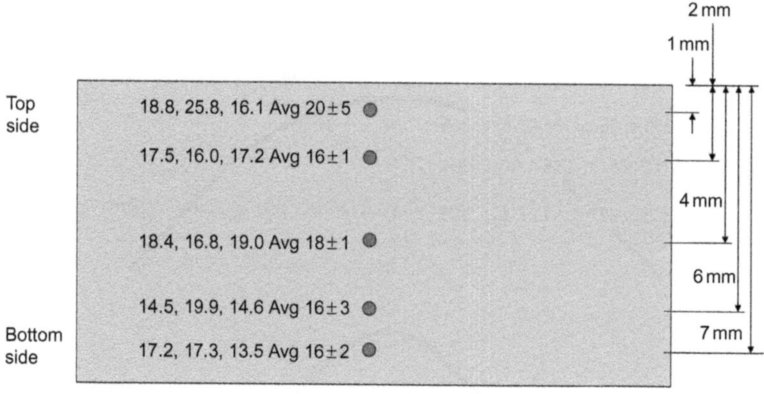


*Figure 4.71 Elongation data in the longitudinal direction for the unformed base material.*

The reduction of area data in the longitudinal direction (parallel to the rolling direction) for the unformed base material is shown in Figure 4.72. By comparing the data with that shown in Section 4.5, it can be seen that there are no significant differences in the reduction of area based on testing direction.

The ultimate strength data in the longitudinal direction (parallel to the rolling direction) for the formed base material is shown in Figure 4.73. All five locations tested for the transverse testing were also tested in the longitudinal direction. It can be seen that there are significant differences in the UTS based on testing direction, by comparing the data below to the data shown in Section 4.5. This suggests

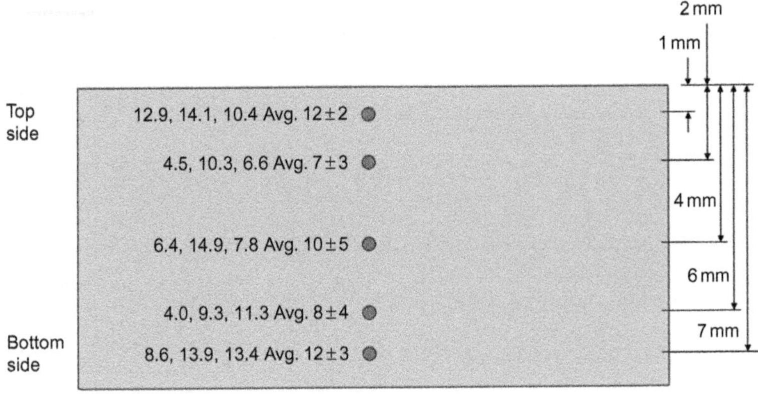


*Figure 4.72 Reduction of area values for specimens tested in the longitudinal direction for the unformed base material.*

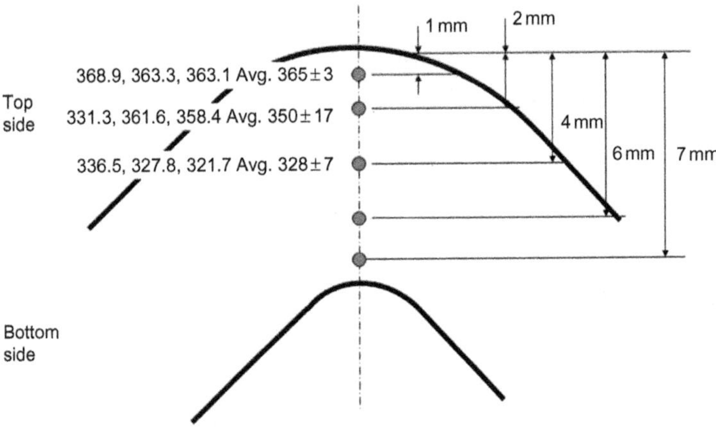


*Figure 4.73 UTS data in the longitudinal direction (parallel to the rolling direction) for the formed base material.*

directionality in the level of work hardening in the transverse and longitudinal directions. The bending operation involves plane–strain condition, as highlighted in Chapter 2. The data suggests that the dislocation substructure developed during bending is dependent of the directional strain, and hence results in anisotropy after forming.

The yield strength data in the longitudinal direction (parallel to the rolling direction) for the formed base material is shown in Figure 4.74. By comparing the data with that shown in Section 4.5, it can be seen that there are significant differences in the yield strength based on

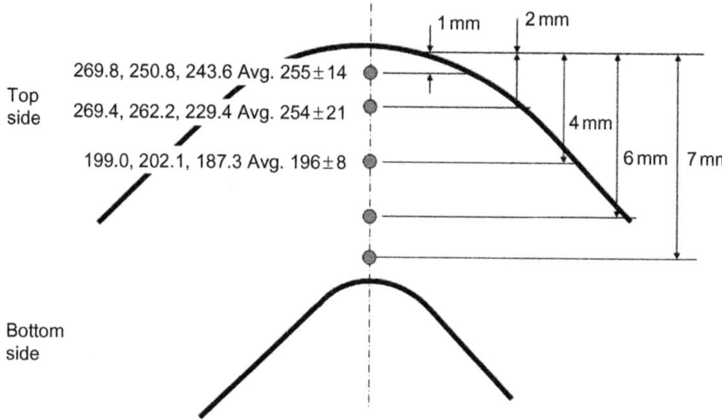


*Figure 4.74 Yield strength data in the longitudinal direction (parallel to the rolling direction) for the formed base material.*

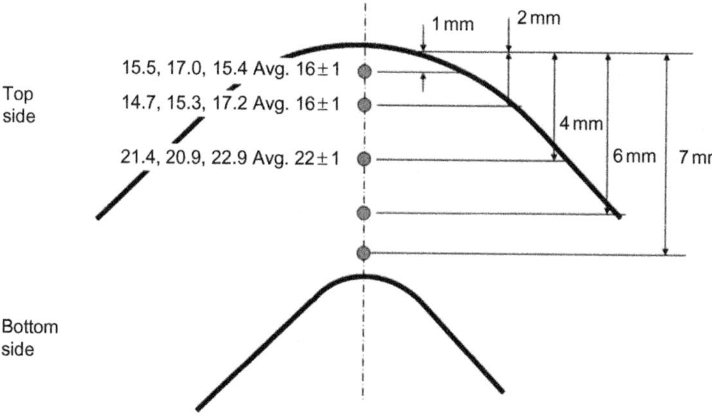


*Figure 4.75 Elongation data in the longitudinal direction (parallel to the rolling direction) for the formed base material.*

testing direction, especially near the surface. This is due to the fact that the work hardening is direction and bending strain dependent as highlighted above.

The elongation data in the longitudinal direction (parallel to rolling direction) for the formed base material is shown in Figure 4.75. By comparing the data with that shown in Section 4.5, it is seen that there is about a 5% difference in elongation versus the transverse testing direction.

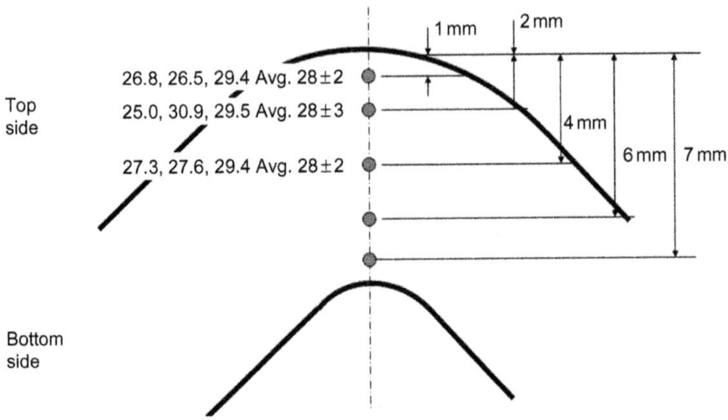


*Figure 4.76 Reduction of area data in the longitudinal direction (parallel to the rolling direction) for the formed base material.*

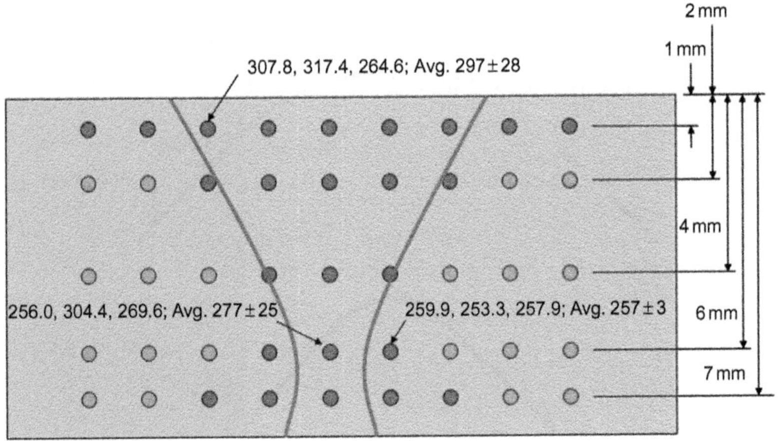


*Figure 4.77 UTS data in the longitudinal direction for the gas metal arc weld samples.*

The reduction of area data in the longitudinal direction (parallel to the rolling direction) for the formed base material is shown in Figure 4.76. By comparing the data with that shown in Section 4.5, it can be seen that there are significant differences in the reduction of area based on testing direction. This is presumably due to the increased elongation and may be related to textural aspects that have not been quantified.

The ultimate strength data in the longitudinal direction for the gas metal arc weld samples is shown in Figure 4.77. Three locations tested

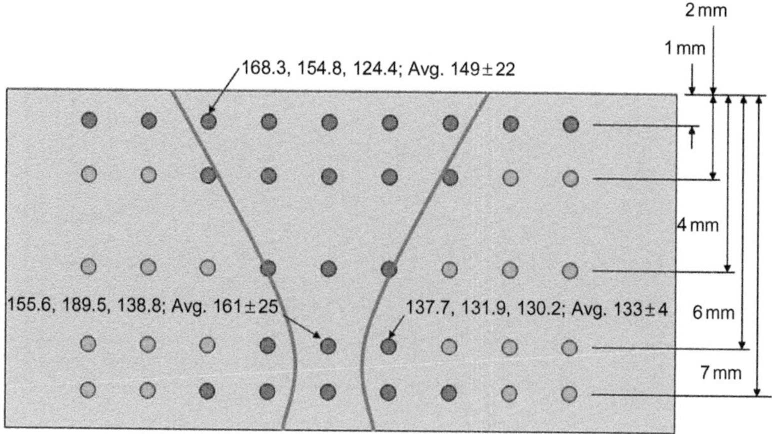


*Figure 4.78 Yield strength data in the longitudinal direction for the gas metal arc weld samples.*

for the transverse testing were also tested in the longitudinal direction. It can be seen that there are no significant differences in the UTS based on testing direction, by comparing the data below to the data shown in Section 4.5.

The yield strength data in the longitudinal direction for the GMAW samples is shown in Figure 4.78. By comparing the data with that shown in Section 4.5, it can be seen that there are no significant differences in the YS based on testing direction.

The elongation data in the longitudinal direction for the gas metal arc weld samples are shown in Figure 4.79. There is no significant difference in elongation based on testing direction, compared with Section 4.5.

The reduction of area data in the longitudinal direction for the GMAW samples are shown in Figure 4.80. By comparing the data with that shown in Section 4.5, it can be seen that there are no significant differences in the reduction of area based on testing direction.

The ultimate strength data in the longitudinal direction for the unformed FSP samples are shown in Figure 4.81. Two locations tested for the transverse testing were also tested in the longitudinal direction. It can be seen that there are no significant differences in the UTS based on testing direction, by comparing the data below to the data shown in Section 4.5.

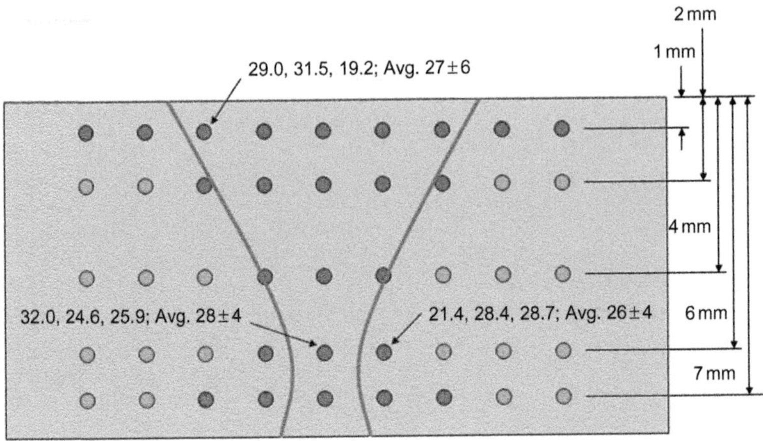


*Figure 4.79 Elongation data in the longitudinal direction for the gas metal arc weld samples.*

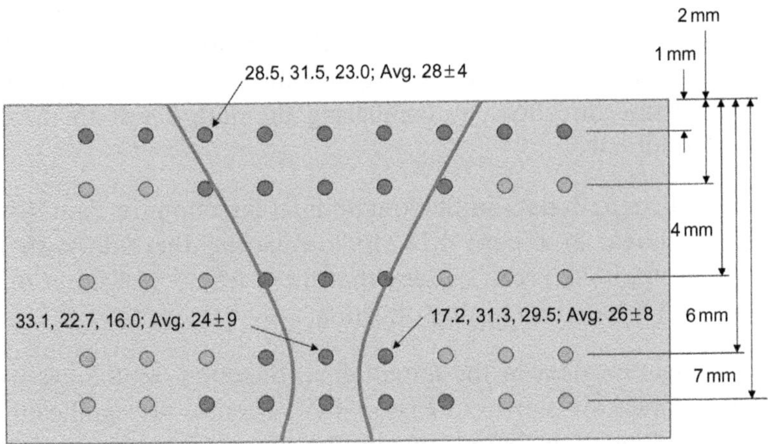


*Figure 4.80 Reduction of area data in the longitudinal direction for the GMAW samples.*

The yield strength data in the longitudinal direction for the unformed FSP samples is shown in Figure 4.82. By comparing the data with that shown in Section 4.5, it can be seen that there are no significant differences in the YS based on testing direction.

The elongation data in the longitudinal direction for the unformed FSP samples is shown in Figure 4.83. By comparing the data with that shown in Section 4.5, it can be seen that there are no significant differences in the elongation based on testing direction.

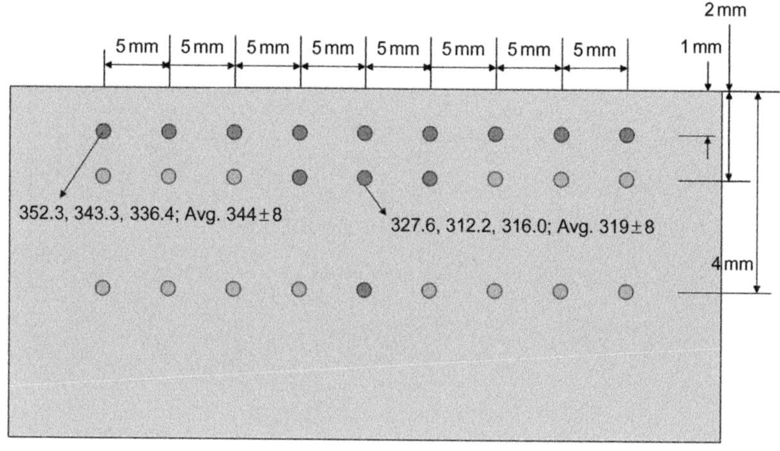


*Figure 4.81 UTS data in the longitudinal direction for the unformed FSP samples.*

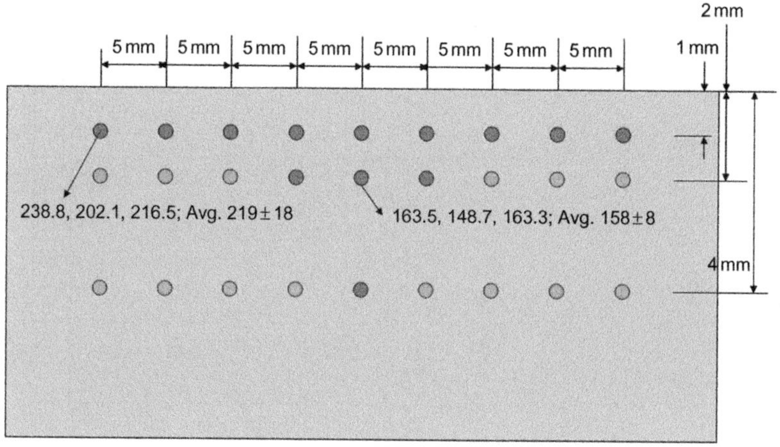


*Figure 4.82 Yield strength data in the longitudinal direction for the unformed FSP samples.*

The reduction of area data in the longitudinal direction for the unformed FSP samples is shown in Figure 4.84. By comparing the data with that shown in Section 4.5, it can be seen that there are no significant differences in the reduction of area based on testing direction.

The ultimate strength data in the longitudinal direction for the formed FSP samples is shown in Figure 4.85. Three locations tested for the transverse testing were also tested in the longitudinal direction.

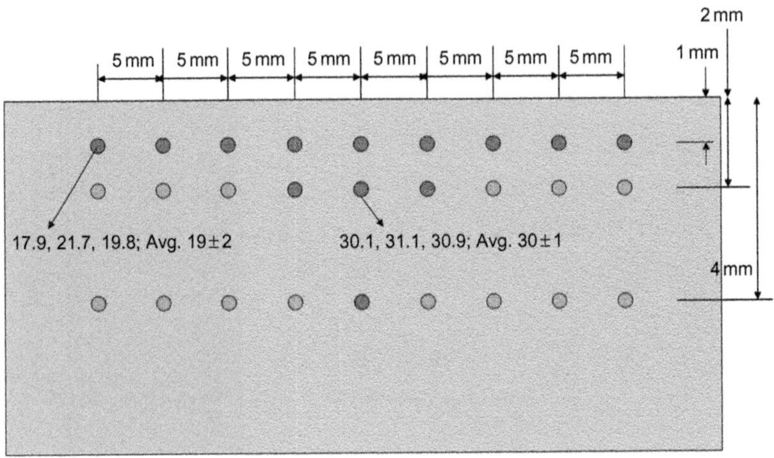


*Figure 4.83 Elongation data in the longitudinal direction for the unformed FSP samples.*

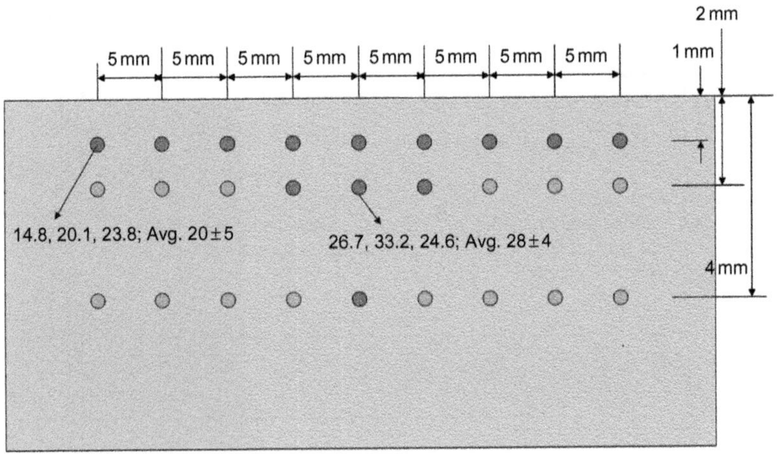


*Figure 4.84 Reduction of area data in the longitudinal direction for the unformed FSP samples.*

It can be seen that there are some differences in the UTS based on testing direction, by comparing the data below to the data shown in Section 4.5. This is presumably due to the strain hardening being preferential based on forming direction as discussed before.

The yield strength data in the longitudinal direction for the formed FSP samples are shown in Figure 4.86. By comparing the data with that shown in Section 4.5, it can be seen that there are significant differences in the YS based on testing direction due to the direction of forming.

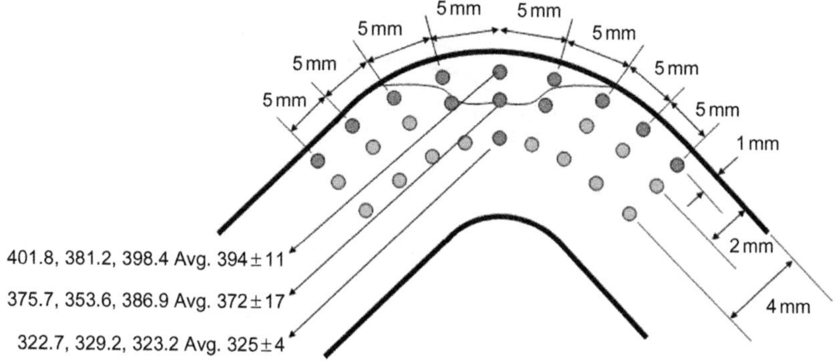


*Figure 4.85 UTS data in the longitudinal direction for the formed FSP samples.*

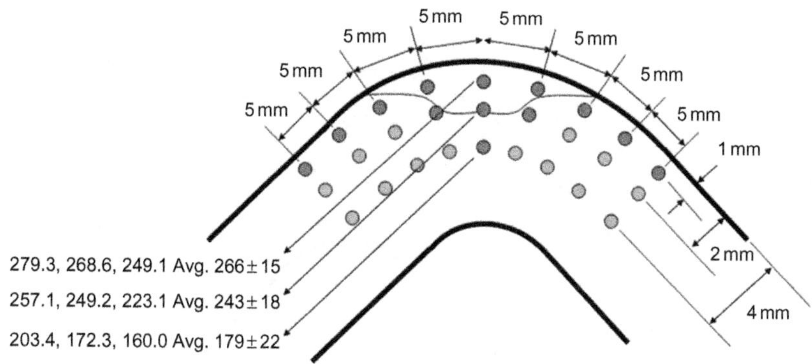


*Figure 4.86 Yield strength data in the longitudinal direction for the formed FSP samples.*

The elongation data in the longitudinal direction for the formed FSP samples are shown in Figure 4.87. By comparing the data with that shown in Section 4.5, it is seen that there are significant differences in the elongation based on testing direction, due to the direction of forming.

The reduction of area data in the longitudinal direction for the formed friction stir processed samples is shown in Figure 4.88. By comparing the data with that shown in Section 4.5, it can be seen that there are significant differences in the reduction of area based on testing direction, due to the difference in elongation. This is a result of the forming direction and associated work hardening.

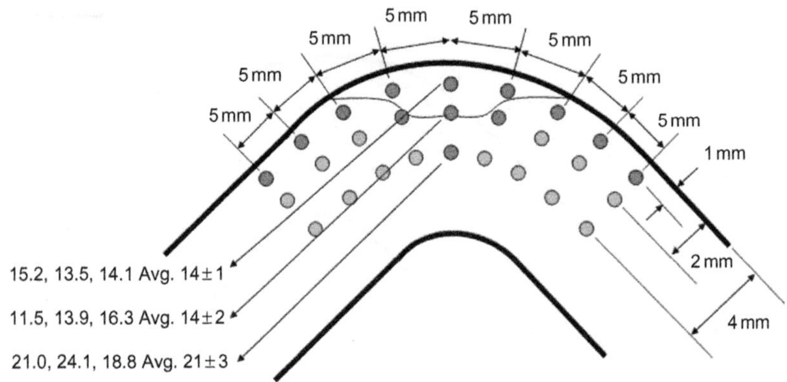


*Figure 4.87 Elongation data in the longitudinal direction for the formed FSP samples.*

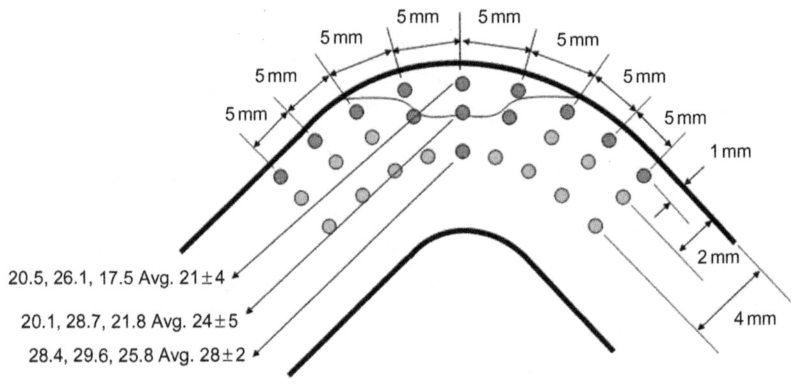


*Figure 4.88 Reduction of area data in the longitudinal direction for the formed friction stir processed samples.*

The ultimate strength, yield strength, and elongation data in the longitudinal direction for the FSW samples are shown in Figure 4.89. It can be seen that there are no significant differences in the UTS based on testing direction, by comparing the data below to the data shown in Section 4.5.

## 4.14 DISTORTION MEASUREMENTS

### 4.14.1 Introduction

Distortion has a significant impact on the structural assembly processes. Excessive distortion causes significant nonvalue-added work. Distorted parts are often processed with a straightening operation to help minimize the effects of the distortion. Distortion is difficult to

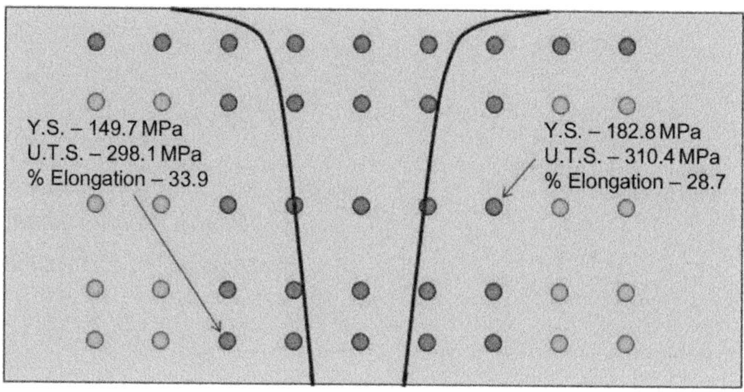


*Figure 4.89 UTS, yield strength, and elongation data in the longitudinal direction for the FSW samples.*

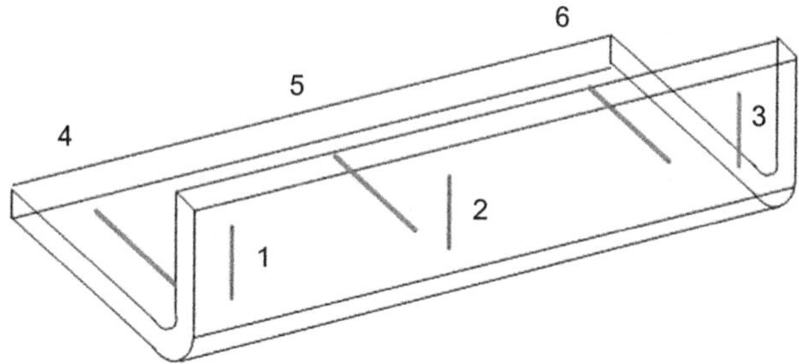


*Figure 4.90 The locations for the angular distortion measurements and the identifying label/number versus location for the subsections.*

completely remove and when not completely removed causes significantly extra effort during the setup for following operations. Thus, understanding the relative distortion effect of friction stir processing with the proposed approach versus the gas metal arc welding used during the traditional approach is important. As such, distortion measurements were taken for each approach to help understand the potential relative effects of each approach.

## 4.14.2 Sample Geometry

To understand the actual distortion, both angular distortion and displacement were measured for both the traditional approach (formed and GMAW spliced) and the proposed approach (FSP and formed). For the traditional approach, the distortion was measured before splicing and after splicing. For the proposed approach, only measurements were taken after FSP and forming, since there was no splicing operations. Figure 4.90 shows the locations for the angular distortion

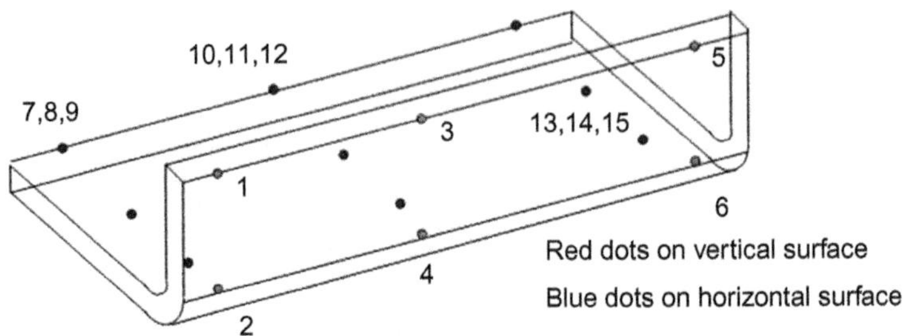


*Figure 4.91 The displacement measurement locations.*

measurements and the identifying label/number versus location for the subsections, or prespliced sections for the traditional approach. Similarly, the displacement measurement locations are shown in Figure 4.91. These prespliced sections are 1800 mm (approximately 6 ft long). The full-length sections were composed of three of the prespliced sections for the traditional approach. They were 5.5 m (18 ft) long. Likewise, the full-length samples (nonspliced) for the proposed approach were also 5.5 m (18 ft) long. For both of these full-length sections, the measurement locations were repeated for each 1.8 m (6 ft), for total number of measurement locations being triple the number of measurement locations of the 1.8 m (6 ft) sections.

For the measurements, the samples were placed on a flat layout table, in the free-state (no clamping, weights, or other means to remove distortion were applied). Angular distortion measurements were made with a digital protractor, with 0.1° resolution. The displacement measurements for the horizontal measurement were made with a height gage and the vertical measurements were made with a square and taper gage and shims, if necessary.

### 4.14.3 Distortion Measurement Results

The full-length stiffeners were fabricated using the traditional approach and the proposed approach. The traditional approach included forming of three 1800 mm (~6 ft) sections to a 1.25″ inner radius, then gas metal arc welding the three sections together. Gas metal arc welding was performed with the WPS that was developed during the initial qualification. All processes match what is currently used on the Littoral Combat Ship. The stiffeners that were fabricated using the traditional approach were first friction stir processed per the WPS developed during the initial qualification along the centerline of the

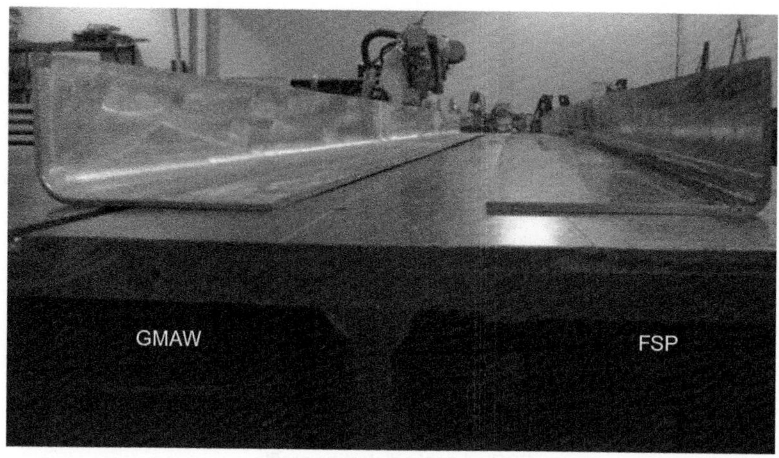


*Figure 4.92 A photograph of one each of the stiffeners produced by GMAW and FSP.*

intended bend axis. The processing was performed on 5400 mm (~18 ft) long strips to match the length of the stiffeners fabricated using the traditional approach. The flat strips were then formed to an 8 mm (5/16″) radius. All material was 8-mm-thick AA 5083-H116 and was formed to create right angle stiffeners with legs of approximately 150 and 250 mm. Three stiffeners for each fabrication approach were created. A photograph of one each of the stiffeners is shown in Figure 4.92.

After completion of the fabrication of the full-length samples, they were nondestructively tested. This included 100% visual inspection, 100% dye-penetrant inspection, and 100% radiographic inspection. Each was inspected per the referenced procedures noted in Section 2. The friction stir processed samples for the proposed approach showed no indications with any of the inspection methods. The gas metal arc weld samples did show some indications, pointing to the relative challenge of the manual GMAW splicing operation used for the traditional manufacturing technique.

Diagrams of the side views of the components to display the general shape of the components are shown in Figure 4.93 and Figure 4.94. It can be seen visually that the proposed approach has more general type distortion versus the wavy distortion of the traditional approach. The stiffeners fabricated with the traditional approach tend to have a single wave between each splice, whereas the proposal approach tends to

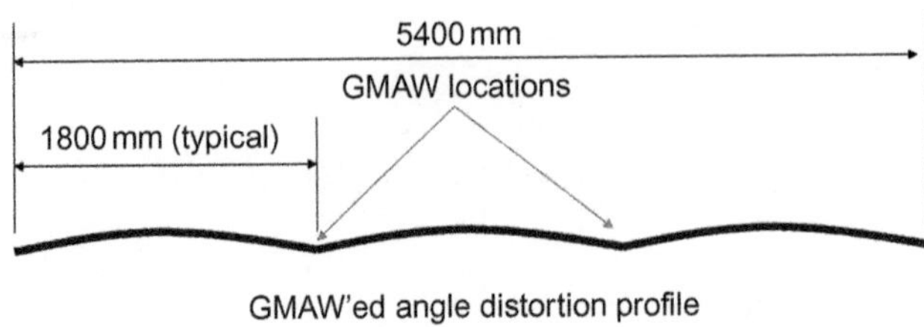


*Figure 4.93 Diagram of the side view of the GMAW component to display the general shape.*

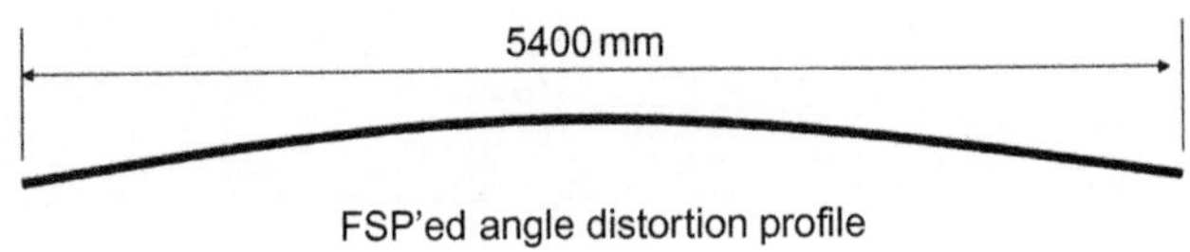


*Figure 4.94 Diagram of the side view of the FSP component to display the general shape.*

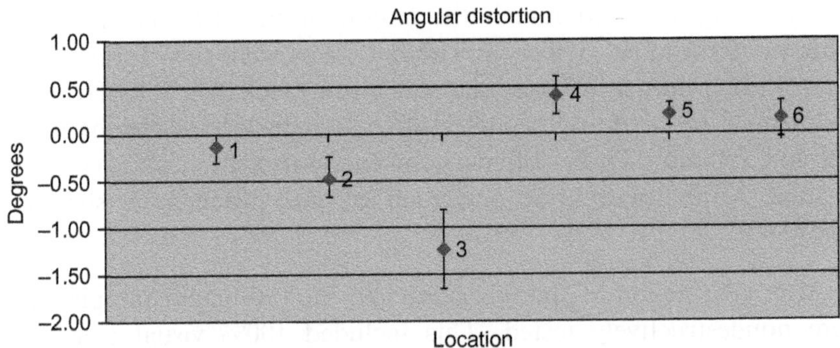


*Figure 4.95 The angular distortion measurements for the prespliced, formed sections with the traditional GMAW approach.*

generate a single larger wave. It is also noted, though not shown, that the distortion of the stiffeners using the traditional approach is not consistent. These differences in distortion have ramifications for fabrication. The single wave distortion of the proposed approach is much easier to manage and eliminate, due to the much lower amount of force required to straighten the part.

The angular distortion measurements for the prespliced, formed sections with the traditional GMAW approach are shown in Figure 4.95. The data are displayed as the average deviation from the nominal

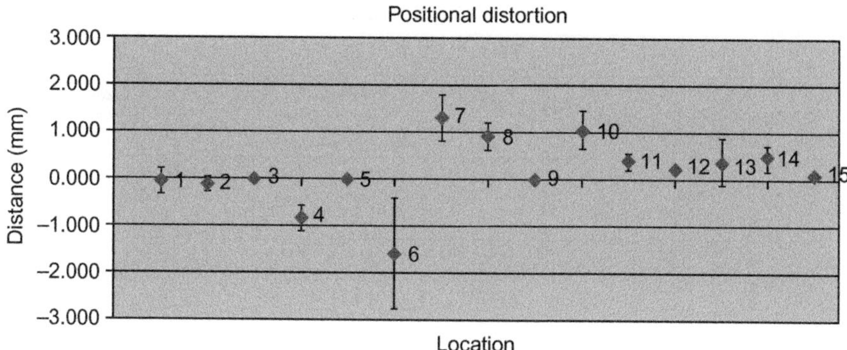


*Figure 4.96 The displacement measurements for the prespliced sections.*

value. For example, for the horizontal leg, the nominal measurement is 0° and for the vertical leg nominal is 90°. Both the average and the standard deviation are shown. The standard deviation is shown by the error bars. The data indicate that the vertical leg and horizontal legs are relatively consistent (low standard deviation), but the horizontal leg gradually flares upward along the length of the prespliced sections, as indicated by locations 1, 2, and 3.

The displacement measurements for the prespliced sections are shown in Figure 4.96. Like the angular displacement measurements, the displacement measurements indicate consistency with a relatively low standard deviation. This indicates the forming operation to generate the large radius for the tradition approach is a consistent and robust process. It is noted that position five has an average and standard deviation of zero, since this is the reference position for the vertical displacement measurements.

Next, the angular distortion measurements are shown in Figure 4.97 for the full-length angles that were fabricated with the traditional GMAW approach. The location data are indicated by the section followed by the measurement location, separated by a comma. For example, the location labeled as "2,3," is the second section measurement location three. The data indicate a similar trend to the prespliced section data. Location "3" tends to have the most angular distortion. In addition, the absolute value is relatively unchanged. However, the standard deviations are now higher, indicating the effect of the gas metal arc welding process is variable.

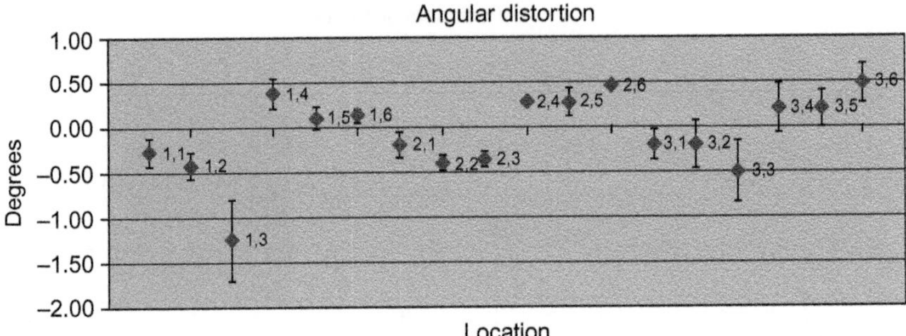


*Figure 4.97 The angular distortion measurements for the full-length angles fabricated with the traditional GMAW approach.*

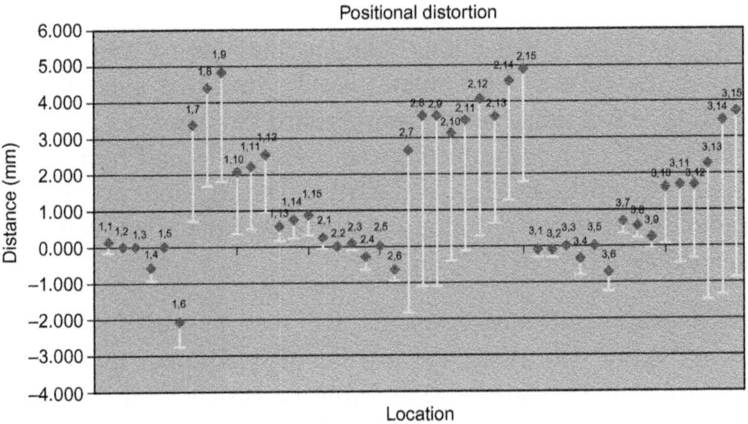


*Figure 4.98 The distortion displacement measurements for the full-length angles fabricated using the traditional approach.*

The distortion displacement measurements for the full-length angles fabricated using the traditional approach are shown in Figure 4.98. In this case, it can be seen that the displacement distortion is significant (average) and is quite variable, as indicated by the standard deviations. It can be seen that the horizontal measurement shows higher variation and average distortion than the vertical leg. This indicates that the part is out of straight in horizontal plane, more than the vertical plane. The data indicate the significant impact of effect of distortion related to gas metal arc welding.

For comparison, the angular distortion measurement data and the displacement data for the new FSP plus bend approach are shown in

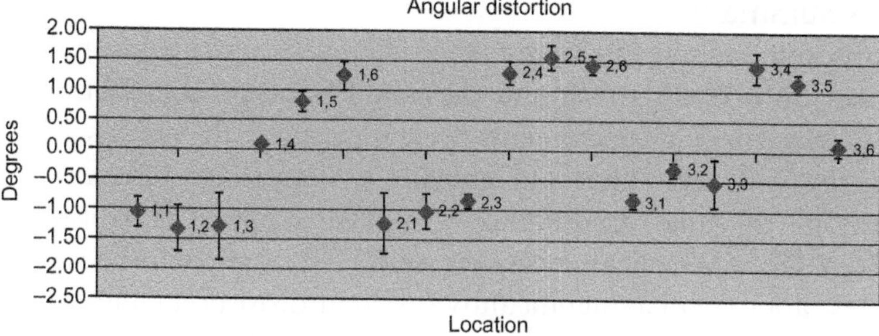


*Figure 4.99 The angular distortion measurement data for the new FSP plus bend approach.*

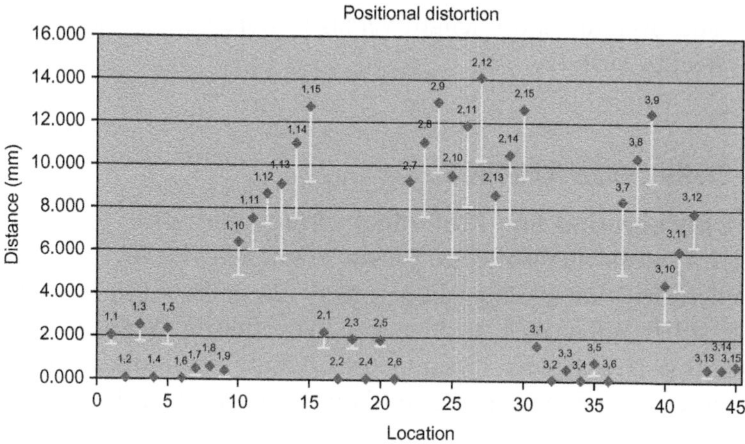


*Figure 4.100 The displacement data for the new FSP plus bend approach.*

Figures 4.99 and 4.100, respectively. The angular distortion data, while referencing the location data shown in the previous figures, indicate that the parts are slightly overbent, but consistent. Further forming process optimization would eliminate this issue. By comparison to the traditional approach, it can be seen that the absolute value or average displacement is higher and displays the single wave trend previously. This difference will have significant benefit for reducing poststiffener fabrication costs. Any straightening operations would be much simpler and setup labor for following assembly operations could be significantly reduced. This could also have a significant impact on the buckling and tripping response.

### 4.14.4 Summary

Distortion measurements were taken on angles fabricated using the traditional GMAW approach and the new FSP approach. Both angular and displacement measurements were measured in various locations along the stiffeners. The data indicate a significant difference in the distortion of the traditional GMAW approach (form short section + GMAW splice) versus the new FSW approach (FSP + form). The new approach has significantly different distortion (single wave vs. multiple waves). This has positive ramifications for the new approach, not only for performance but also for the production costs. With the new approach, nonvalue-added operations of straightening and excessive setup can be significantly reduced because the distortion profile is broad and over the entire length versus a more wrinkle profile of the GMAW approach.

## 4.15 CORROSION TESTING

### 4.15.1 Background and Preliminary Investigation

#### 4.15.1.1 Typical Microstructure Parent Metal 5083-H116

The microstructure of the parent metal AA 5083-H116 is shown in Figure 4.101, after etch with a 5% hydrofluoric acid solution. As shown, this is a typical rolled microstructure resulting in elongated high aspect ratio grains. The grain boundaries are revealed because of the presence of β phase. Intergranular and grain boundary precipitates, identified as β phase, $Mg_3Al_2$, are the finer precipitates. The relatively larger particles are iron-rich constituent particles.

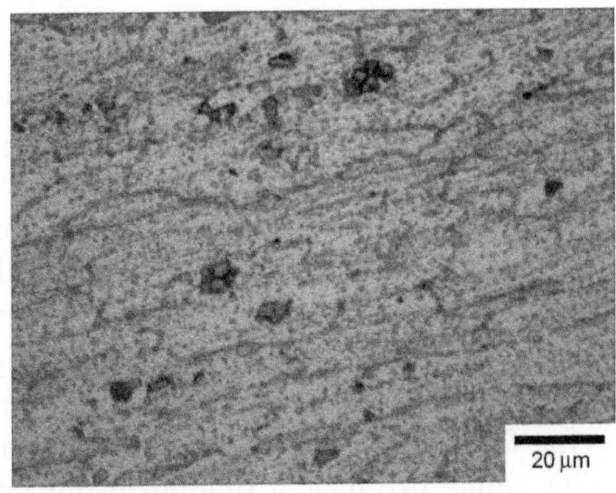


*Figure 4.101 The microstructure of the parent metal AA 5083-H116.*

The temper designation H-11 applies to products that incur sufficient strain hardening after the final anneal that they fail to qualify as annealed but not so much or so consistent an amount of strain hardening that they qualify as H-1. Also, the H-116 temper is used for high magnesium alloys, such as AA 5083, and involves special temperature control during fabrication to achieve a microstructural pattern of precipitates that increases the resistance of the alloy to intergranular corrosion (IGC) and stress corrosion cracking (SCC).

#### 4.15.1.2 Microstructure and Property Changes Due to FSP + Room Temperature Bending

##### 4.15.1.2.1 Composite Microstructure

Following partial penetration FSP, there are three zones in the 5083-H116 alloy (Figure 4.102) where the parent metal microstructure has changed including (1) the HAZ, where the metal experiences elevated temperature for a short time but no deformation is induced (location 1), (2) the thermomechanically affected zone (TMAZ) where the metal experiences both heat and deformation but the combination is insufficient for recrystallization (locations 2 and 3_2), and (3) the stir zone (STIR ZONE) where the metal experiences sufficient heat and deformation to facilitate complete recrystallization creating a fine grain equiaxed microstructure (location 3_1).

Representative micrographs of these different microstructural zones are illustrated in Figure 4.103. To reveal the grain boundaries, the specimens were sensitized by aging at 160°C for 24 h. The sensitization treatment results in additional precipitation of β phase on the grain boundaries. For comparison, Figure 4.103C, a micrograph of the base material and Figure 4.103E, a micrograph of the stir zone

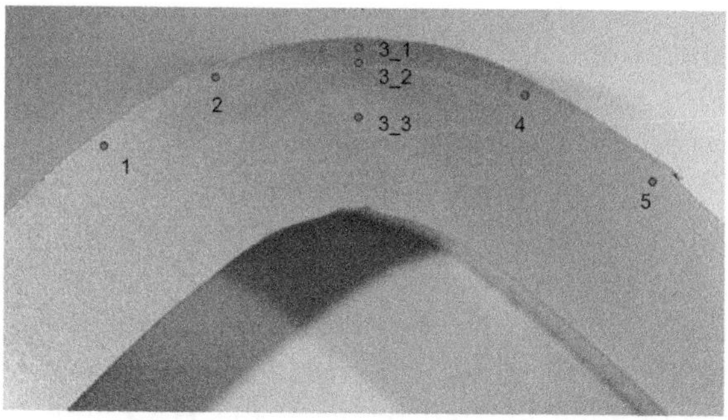


*Figure 4.102 Following partial penetration FSP, there are three zones in the 5083-H116 alloy.*

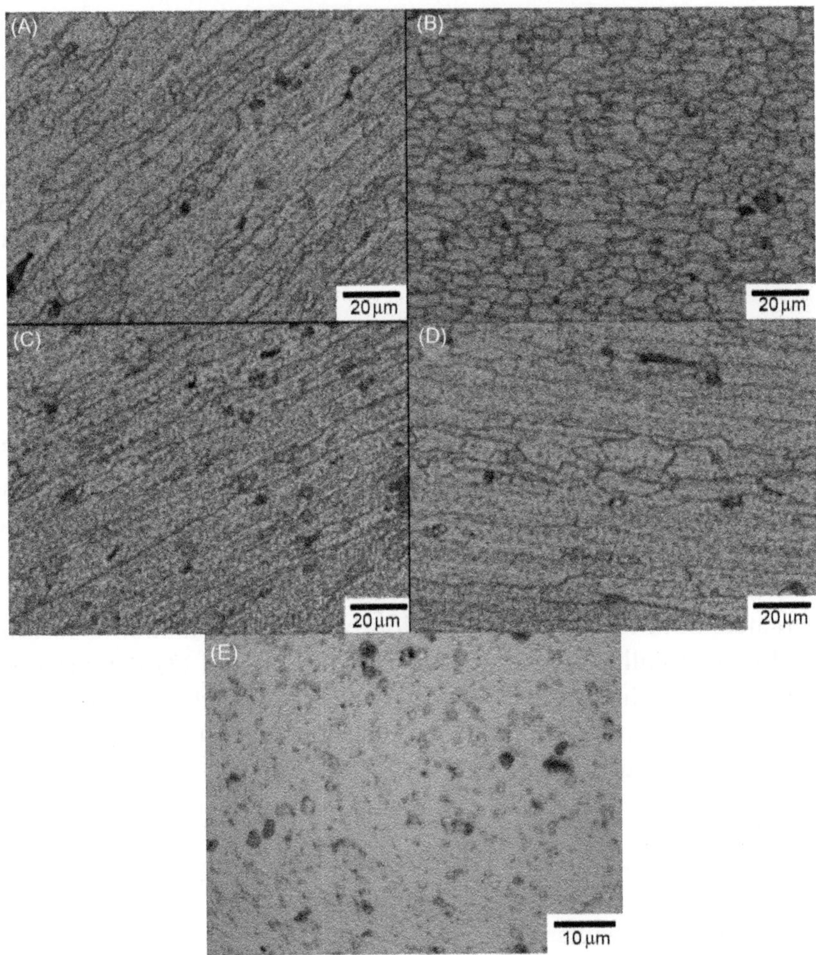


*Figure 4.103 Representative micrographs of these different microstructural zones.*

without sensitization are also included. The HAZ, shown in Figure 4.103D is similar to that created by conventional fusion welding techniques except the HAZ created by FSP is generally smaller due to the reduced heat input (this is friction stir process parameter dependent). Also, since the FSP is partial penetration (<25% of the sheet thickness), the HAZ exists both adjacent to the TMAZ, shown in Figure 4.103A, and below the stir zone, shown in Figure 4.103B. The HAZ and parent metal microstructures look the same. The TMAZ is a small zone located between the HAZ and the stir zone. Due to the FSP tool design, that is, little deformation below the tool pin, if there is a TMAZ below the stir zone, it is relatively small. The TMAZ is differentiated from the HAZ by a small amount of deformation, but it is difficult to quantitatively access if any work hardening remains due to the coincident thermal cycle, for example, the dislocation density is reduced by annealing. Most likely, the strain hardening associated with the subsequent bending would be significantly greater than any strain hardening associated with FSP, albeit this is only speculation. The stir zone experiences the greatest microstructural change whereby a fine grain equiaxed microstructure is created by recrystallization. In addition, as can be noted from the micrograph without sensitization does not show grain boundaries. The implication is that the β phase is mostly dissolved in the stir zone. The iron containing constituent particles are also refined and uniformly distributed. These factors are likely to result in lower corrosion susceptibility.

#### 4.15.1.2.2 Hardness

Following FSP, hardness in the stir zone in aluminum alloys is most often reduced due to precipitate dissolution or annealing of any strain hardening. However, for the case of 5083-H116, the reverse has been shown by other investigators, that is, the stir zone becomes harder than the surrounding HAZ. Kumagai and Tanaka (1999) showed a slight increase in hardness across the stir zone compared to the HAZ for a slightly hardened 5083-H112 but the increase was less than 6%. More recently, work by Vilaca et al. (submitted for publication) duplicated these same results illustrating an 8% increase in hardness in the stir zone over the parent metal. This hardness increase may be due to the very fine grain size in the stir zone. Yield strength can be directly proportional to hardness as shown by Sato and Kokawa (2001), albeit for a AA 6063. This increased hardness can be important for fatigue life as fatigue life increases for increasing yield strength.

#### 4.15.1.2.3 Precipitate Distribution/Dissolution

In studies by Pao et al. (2005) on FSW 5456-H116 (a higher Mg content than 5083), a completely different precipitate distribution is observed in the weld nugget. This likely indicates that the temperature achieved during welding was sufficient to dissolve the former precipitate distributions and permit formation of this different precipitate during cooling.

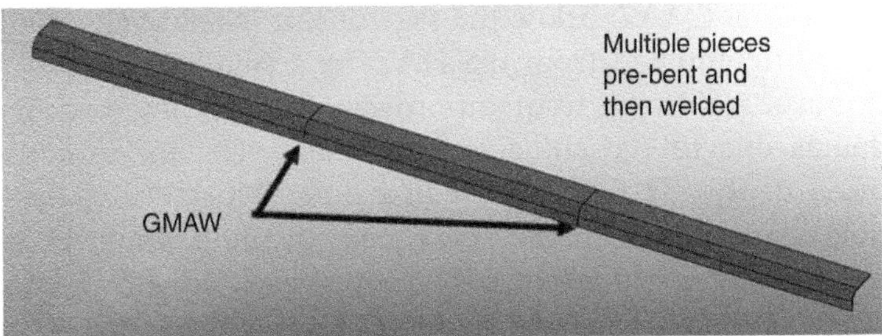


*Figure 4.104 The fabrication sequence for the current approach for angle fabrication involves bending of short sections to right angles followed by GMAW of the short sections, referred to as splicing.*

#### 4.15.1.2.4 Cold Work Distribution Due to FSP/Bending

As discussed in detail in Chapter 1, the fabrication sequence for the current approach for angle fabrication involves bending of short sections to right angles followed by GMAW of the short sections, referred to as splicing, Figure 4.104. Thus, the sheet is work hardened at the bend for most of the angle length, but at the fusion-welded splices, all work hardening from bending is lost in the fusion weld and most is lost in the adjacent HAZ. Further, in the weld area, any work hardening introduced into the plate prior to bending (H-11 temper) will also be lost. Thus, the current angle fabrication practice produces an angle with varying work hardening and thus varied properties along the length.

The proposed FSP/bending approach to fabricate angles uses a different sequence of metal working operations. First, in the flat condition, the sheet is friction stir processed along the length. The heat and deformation in the FSP zone will partially anneal any prior work hardening resulting in a lower dislocation density (Williams et al., 2006). However, as discussed above, hardness in the FSP zone for the H116 temper still increases. Following FSP, the sheet is bent to a right angle with the FSP zone in tension. Thus, work hardening is reintroduced along the angle length creating a structure with homogeneous properties along the length. The magnitude and extent (width across the sheet) of work hardening on the tensile surface has not been established. This can be established experimentally by etching a grid pattern across the FSP zone prior to bending and subsequently measuring the grid dimensions or by a simple modeling approach. Thus, the majority volume of the angle will retain parent metal properties while the FSP/bent zone will contain additional work hardening, as noted in Section 4.2.

4.15.1.2.5 Residual Stresses

The magnitude and distribution of residual stresses will be different following FSP/bending versus bending/GMAW. Residual stresses can influence mechanical, fatigue, and corrosion properties. Thus, it is valuable to be aware of the residual stress state. In general, residual stresses associated with FSP are relatively low, that is, considerably less than the yield strength. Unfortunately, residual stress data on AA 5083 have not been available. However, residual stress measurements by Williams et al. (2006) for AA 2024-T3 are typical of residual stress data available in the literature. Transverse residual stresses are low (<20 MPa) and are in compression. The highest tensile residual stresses are in the transition region between the stir zone and TMAZ but are still <130 MPa. Within the stir zone, the longitudinal residual stresses are less (~100 MPa). Thus, residual stresses never exceeded half of the yield strength of the alloy, considerably below that often created by fusion welding processes. Other investigators have shown similar results for other aluminum alloys. For this application, partial penetration FSP + bending will create a complex residual stress pattern. In order to map residual stresses completely (at least in the longitudinal orientation), the contour method would need to be used to map the residual stresses for the entire cross section of the angle (Prime, 2001). In addition, distortion due to FSP is far less than that introduced by gas metal arc welding (see Section 4.7), suggesting the proposed FSP/bending process will have lower residual stresses.

#### 4.15.1.3 Corrosion Mechanisms in Aluminum Alloys with Emphasis on AA 5083

4.15.1.3.1 Intergranular

IGC is preferential attack close to grain boundaries, which can cause severe damage with a relatively small amount of corrosion. This is caused by different levels of corrosion resistance between the bulk grain and areas near to, or at, the grain boundary. In aluminum alloys, the difference in corrosion resistance can be due to precipitate formation or redistribution along the grain boundary. Precipitates that are more anodic or cathodic than the surrounding area will either preferentially corrode or encourage the surrounding area to preferentially corrode. In either case, macroscopic preferential grain boundary attack will occur. Generally, age-hardened alloys are more susceptible to IGC than the solid solution-strengthened aluminum alloys (such as the 5XXX Al alloy series) due to the nature of the former's strengthening mechanism involving distributions of fine precipitates. However, additional thermal cycles, such as those experienced in welding and preheating, might also develop a susceptible microstructure in a solid solution-strengthened aluminum alloy.

For AA 5083, the precipitates that might decorate grain boundaries are $Al_3Mg_2$ and $Al_6Mn$. However, as discussed in Section 4.15.1.3.4, a considerably long time at an intermediate temperature would be required for $Al_3Mg_2$ to substantially decorate grain boundaries. Further, in the work by Searles et al. (2001), even when there was a continuous $Al_3Mg_2$ grain boundary phase (189 h at 150°C), there was essentially no Mg depletion adjacent to the grain boundaries. Thus, the short time at temperature associated with FSP would not create a microstructure susceptible to intergranular attack. Further, the Mn content in AA 5083 is low (~0.7%) and there have been no reports of the $Al_6Mn$ precipitate associated with IGC.

#### 4.15.1.3.2 Crevice

When two surfaces touch, a narrow gap or crevice forms and when an electrolyte is present, crevice corrosion can occur due to the differential aeration cell setup between the mouth and tip of the crevice. A very tight or thin crevice is most likely to promote corrosion, as the diffusion of oxygen is particularly limited creating a greater difference in oxygen concentration. Crevice corrosion does not have a definitive morphology, as it very much depends on the shape and size of the crevice and the wetted area. It often appears as localized corrosion in the form of shallow pits which eventually join together into an area of general metal loss. Crevice corrosion is particularly important in lap and fillet welded joints, or partial penetration butt welds where crevices cannot be avoided. If an aluminum alloy angle is in close proximity to an alloy of a different potential, then crevice corrosion might be expected with or without FSP.

#### 4.15.1.3.3 Pitting

Pitting is localized corrosion resulting from the breakdown of the surface oxide film. Pits occur as discrete areas of metal loss in varying morphologies; from steep sided and narrow to shallow and wide shapes. In the case of aluminum alloys, pits are typically narrow and steep sided and often found in clusters. Pitting requires moisture, oxygen, and an aggressive species such as chloride to be present for initiation. Pits often initiate next to or at inclusions and intermetallic particles in a localized corrosion cell. If inclusions and intermetallic particles are more cathodic than the surrounding matrix, corrosion will preferentially occur in the surrounding matrix, but if particles are more anodic than the surrounding matrix, they may preferentially corrode instead. Once initiated, pits propagate by autocatalysis, in which the corrosion processes inside the pit generate an acidic environment that prevents repassivation.

In general, the 5XXX series alloys are less susceptible than the higher strength 2XXX and 7XXX series alloys. Although the work of Searles et al. (2001) was directed to SCC, they did comment on pitting in highly sensitized AA 5083, that is, samples aged for 189 h at 150°C. "Pitting on the polished samples of surfaces of the tensile samples was minimal, ...." Thus, even when grain boundaries were highly decorated with β phase, pitting was minimal. Further, if there are large preexisting second phase particles, FSP of AA 5083 has been shown to break up and redistribute these particles both in the TMAZ and stir zone (Vitek et al., 1998). Additional pitting corrosion studies were completed on FSW 5083-H32 by Zucchi et al. (2001). These investigators confirmed the good pitting resistance of the parent metal AA 5083. However, pits were shown to be more numerous on the base alloy than on the weld in the FSW samples. Thus, qualitatively, the FSW nugget showed even greater pitting resistance than the base alloy. The higher pitting resistance of the FSW nugget than that of the base alloy was confirmed by recording the anodic and cathodic polarization curves on electrodes of the base alloy and the FSW nugget. The pitting potential of the FSW was more positive by approximately 10 mV compared to that of the base alloy (Zucchi et al., 2001). However, since the error of this type of testing is ±40 mV, there is essentially no difference between base metal and the FSW nugget according to this study.

#### 4.15.1.3.4 Stress Corrosion Cracking

SCC occurs when a specific alloy is exposed to a particular corrosion environment under stress. The stress can be either applied or a residual stress from prior secondary processing operations, such as bending or friction stir processing. In aluminum alloys, susceptibility to SCC may occur in chloride-containing aqueous environments. However, in extreme cases, even moist air can cause SCC. The microstructure of the aluminum alloy is important in determining its susceptibility to SCC, since mechanisms of SCC involve mostly intergranular type of attack.

Specifically, aluminum alloys that contain appreciable amounts of soluble alloying elements, primarily, copper, magnesium, silicon, and zinc, are susceptible to SCC. This includes AA 5083 with a nominal magnesium content of 4.5%. However, the microstructure dictates the degree of susceptibility (if any). In work by Searles et al. (2001), the SCC susceptibility was shown to be associated with a high rate of dissolution of the electrochemically active β phase, $Al_3Mg_2$, which is precipitated on grain boundaries. Thus, this grain boundary precipitate must be present for SCC to occur in AA 5083. The β phase precipitates on grain boundaries only if the alloy is subjected to temperatures ranging from 50°C to 200°C for sufficiently long periods of time (e.g., >80 h at 150°C). During FSP, temperatures are considerably higher (~500°C) and times considerably shorter (approximately a few seconds) than the sensitization envelope. Thus, FSP would not be expected to create a microstructure any more susceptible to SCC than the parent metal itself. In fact, at the elevated FSP temperatures, any preexisting β phase would likely go into solution resulting in a microstructure in the FSP zone less susceptible to SCC. In the far HAZ, where the temperatures are lower and within the sensitization envelope, the times are still extremely short and the β phase should not precipitate at grain boundaries. Additional SCC studies were performed on friction stir welded 5083-H32 by Zucchi et al. (2001). SCC tests were performed in both the EXCO solution (4 M sodium chloride + 0.5 M potassium nitrate + 0.1 M nitric acid) and a 3.5% sodium chloride + 0.3 g/L $H_2O_2$. Results showed the friction stir weld to not be susceptible to SCC in the chloride solution. In the EXCO solution, the friction stir weld was more corrosion resistant than the base alloy. Further, work or strain hardening is a requirement for SSC in AA 5083. For example, the microstructure of 5083-O is not susceptible to SCC. It is unlikely that the H-116 temper, with a very low level of work hardening is susceptible to SCC. Further, the small amount of work introduced by bending is not likely to alter this conclusion.

#### 4.15.1.3.5 Exfoliation

Exfoliation corrosion is a type of subsurface attack usually found in thin, heavily worked materials. The microstructural constituents, remnant segregation, and grains have a planar distribution in the product working plane and this leads to lamellar attack (exfoliation) parallel to the metal surface. The result of exfoliation corrosion is that thin layers have relative lack of corrosion are detached from the bulk between corroded regions, in a lamellar fashion, due to forces created by the volume of the corrosion product. Exfoliation corrosion tests were performed by Vilica et al. (submitted for publication), on FSW 5083-H111 using the standard ASTM G66 procedure (ASTM G 66, 1999). Results in the parent metal verified exfoliation corrosion in the parent metal in the form of homogeneously distributed pitting. As the HAZ was approached in the FSW sample, the exfoliation corrosion was strongly reduced. Similarly, there was less exfoliation corrosion reported in the stir zone. However, where flash was not removed, there was a concentration of corrosion pits within the parent metal. This is likely due to the concentration of the corrosion environment at the flash illustrating the need for flash removal. Similar exfoliation corrosion results were obtained by Reynolds (1998) in FSW AA 5454-O using the ASSET test. All of the FSW AA 5454-O test specimens exhibited an appearance corresponding to the *N* rating for "no appreciable attack."

#### 4.15.1.3.6 Galvanic Corrosion

Galvanic corrosion is a condition caused as a result of a conducting liquid making contact with two different metals which are not properly isolated physically and/or electrically. Engineers have long been aware of this issue especially when Al and Fe alloys are in close proximity. The corrosion problems resulting from the use of galvanically dissimilar metals is further complicated by a crevice between components. Explosion-welded aluminum to steel transition joints was a solution used to eliminate crevices. However, this does not eliminate the galvanic-driven corrosion but does create a product that is easily protected by conventional corrosion management techniques. Friction stir processing does not alter the potential difference between Al and steel alloys in the galvanic series in seawater. The only change that might have some influence is a change in the surface oxide. However, FSP samples are lightly machined following FSP creating a surface that rapidly repassivates, that is, the surface oxide readily and rapidly reforms. In areas experiencing elevated temperature from the FSP but no machining, a more robust oxide might even be expected. Thus, conventional corrosion management techniques should still apply. Unfortunately, there are no experimental results to support this discussion in the literature.

#### 4.15.1.4 How Microstructural Changes Alter Corrosion Sensitivity in FSP 5083-H116 Aluminum

It is difficult to obtain corrosion data for the specific alloy, temper, and secondary processing used to fabricate angles via FSP/bending. Thus, eventually it will be necessary to perform all the appropriate tests for the different corrosion types using angles fabricate via FSP/bending. Further, accelerated corrosion tests only provide an estimate of corrosion sensitivity. The best measure of corrosion resistance is in situ exposure and subsequent evaluation of structures installed on a representative application. In situ tests are planned and will be performed in below-deck locations and an application will be installed for a noncritical application in an external/exposed environment. However, based on the microstructural changes associated with FSP/bending, published data, and indirect data, it is reasonable to hypothesize that the corrosion resistance following FSP/bending will be equivalent to or better than that for GMAW angles and for most/all corrosion types, even superior to the parent metal corrosion resistance.

Pertinent information from publications above and all research leads to the conclusion that FSP should not adversely affect the corrosion behavior of 5083-H116 (Prime, 2001; Vitek et al., 1998; Zucchi et al., 2001; ASTM G 66, 1999; Reynolds, 1998). Unfortunately, not all data were available to the public and not all information was published. Thus, until direct and specific corrosion tests were performed for specific conditions, engineers must rely on observations from others and on our knowledge of how different microstructures respond to corrosive environments. For example, Scandinavian aluminum extruders were the first to commercially apply FSW for the manufacture of panels for ship decks (Second EuroStir Workshop, 2002). It is believed that no specific problems of corrosion of FSW weldments have been reported in alloys of interest to the shipping sector to date (Stuart Bond, 2003). Further, since melting and resolidification that typically occur in the traditional arc welding processes are eliminated, friction stir welds do not exhibit the severe microstructure changes associated with fusion welds. In addition, no filler metal is used so the weld has the same composition as that of the parent metal.

##### 4.15.1.4.1 Intergranular

IGC requires a precipitate phase at the grain boundaries and/or a depletion of the solute adjacent to the grain boundaries. FSP exposes the AA 5083 to a short time at high temperature in the FSP zone resulting in a fine grain, fully recrystallized, equiaxed microstructure. There is insufficient time at temperature for a second phase to precipitate at grain boundaries in the FSP zone. In regions adjacent to the FSP zone, there is a gradient in temperature decreasing from ~500°C in the FSP zone to room temperature in the far HAZ. Times at temperature are short (<1 min.) and are insufficient to influence the grain boundary precipitate morphology. Thus, based on the FSP microstructure, there is no reason to believe that FSP would reduce the IGC resistance of 5083-H116.

#### 4.15.1.4.2 Crevice

Crevice corrosion is geometry dependent and FSP does not change the alloy's galvanic potential. Thus, for the same structure, there is no reason to suspect that the microstructure resulting from FSP would be any more susceptible to crevice corrosion than the parent metal or GMAW microstructures.

#### 4.15.1.4.3 Pitting

Pitting corrosion is associated with dissolution of second phase particles or the zone adjacent to second phase particles. There is no evidence to suggest that FSP creates additional second phase particles or coarsens second phase particles in AA 5083. Conversely, if large second phase particles exist in the parent metal, likely they would be fractured and redistributed following FSP resulting in reduced susceptibility to pitting corrosion (Vitek et al., 1998). Evidence has been presented to show that pitting corrosion resistance of FSP AA 5083 is superior to that of the parent metal (Zucchi et al., 2001).

#### 4.15.1.4.4 Stress Corrosion

There is considerable evidence identifying $Al_2Mg_3$ as the grain boundary phase associated with SCC in AA 5083 (Prime, 2001). Based on our experience with FSP of AA 5083, the short time, high temperature FSP procedure does not meet the time/temperature criteria for creating a continuous or even a discontinuous $Al_2Mg_3$ grain boundary precipitate. Further, depending on the environment, evidence is available in the literature showing equivalent or even superior SCC resistance of FSP AA 5083 to that of the parent metal (Vitek et al., 1998).

#### 4.15.1.4.5 Exfoliation

Exfoliation corrosion is associated with attack of the high aspect ratio, long planar grain boundaries in aluminum alloys following a secondary metal processing procedure such as rolling that introduces a high level of work. First, FSP creates a very fine fully recrystallized equiaxed microstructure. Fine grain microstructures are more resistant to exfoliation corrosion than the planar elongated parent metal microstructure. Second, any prior work hardening is eliminated or significantly reduced following FSP. Based on the FSP microstructure, exfoliation corrosion following FSP is superior to that of the parent metal microstructure.

#### 4.15.1.4.6 Galvanic Corrosion

There are no data comparing the galvanic potential difference between parent metal 5083-H116 and that of FSP 5083-H116. Thus, it is only speculation that no significant difference would be created by FSP. However, regardless of any galvanic potential differences, if AA 5083 is used in contact with alloys with a different galvanic potential, conventional corrosion resistant measures should be employed.

## 4.15.2 SCC Testing Results

Stress corrosion tests have been performed per ASTM G44 (ASTM G 49, 1985). SCC by alternate immersion in 3.5 wt% sodium chloride solution as specified in ASTM G44 was conducted for sample types A (unformed base material), B (unformed friction stir processed material using scroll shouldered tool), C (friction stir welded), D (formed friction stir processed material), E (gas metal arc welded), F (formed base material), and I (unformed friction stir processed material using stepped spiral tool). Three samples each from all the sample types were subjected to alternate immersion in the solution for hourly cycle of 10 min immersed in the solution and 50 min out of the solution. This cycle was continued for 28 days and after 28 days macroscopic and microscopic examination of the surfaces was carried out in order to estimate the SCC susceptibility.

### 4.15.2.1 SCC Testing Summary

As mentioned in ASTM G44, the material showing continuous crack originating from the surface penetrating into it should be classified under susceptible category and material showing pit formation on the surface should not be classified under susceptible category. After 28 days period surface observation revealed no continuous cracking for

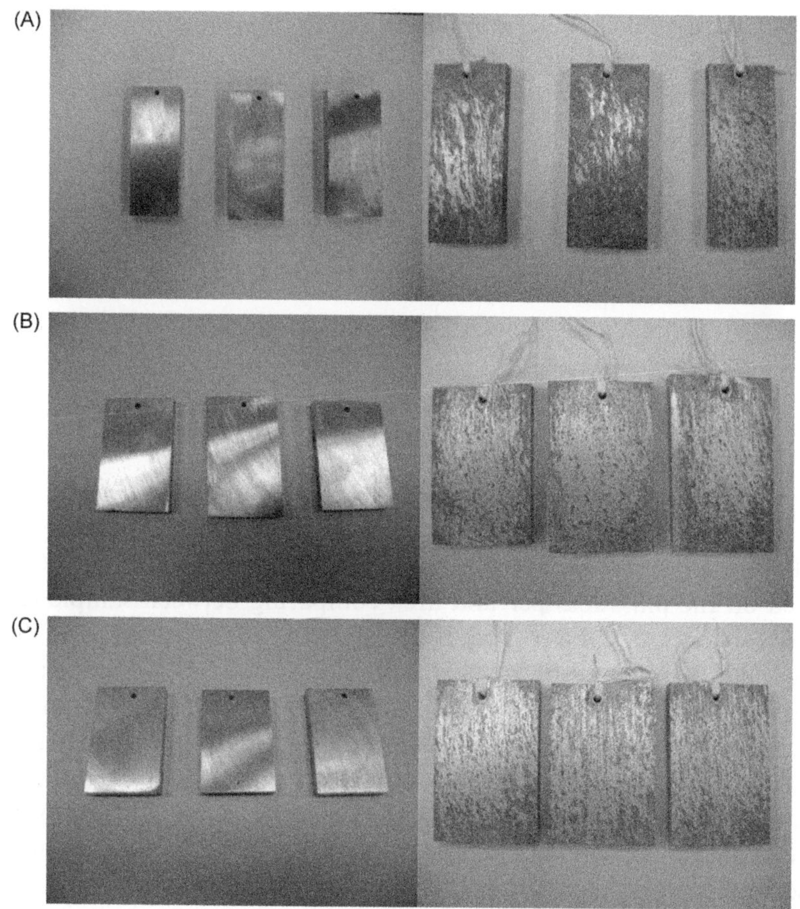

*Figure 4.105 Various samples before and after testing according to ASTM G44. (A) Base material specimens, (B) FSW specimens, and (C) GMAW specimens.*

any of the conditions. Thus, it can be concluded that none of the alloy conditions showed SCC susceptibility even though excessive pit formation was observed over the surface. Various samples before and after testing are shown Figure 4.105A–C.

### 4.15.3 IGC Testing Results

IGC testing was performed per ASTM G67-04 (ASTM G 67, 2004). IGC susceptibility of 5083-H111 base material, gas metal arc welded material, friction stir processed material and friction stir welded material was estimated using ASTM G67-04. Both formed and unformed samples were also tested, where applicable, the samples were exposed to concentrated nitric acid solution for 24 h at 30°C.

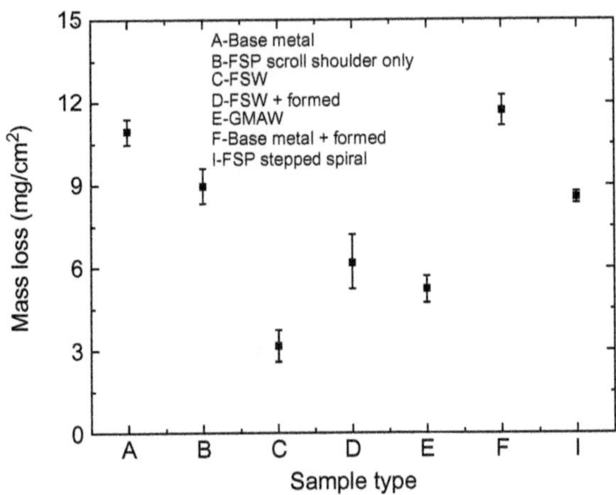


*Figure 4.106 Mass loss per unit area for different sample types.*

Figure 4.106 shows mass loss per unit area for different sample types. Based on the mass loss per unit area comparing between sample types A-base material, C-FSW and E-GMAW, it was observed that base material, GMAW and FSW showed resistance for IGC, with ranking of base material < GMAW < FSW. Comparing between type A-base material and type F-base metal + forming, it was observed that bending, that is, imparting plastic deformation to material, caused to marginally deteriorate the corrosion behavior with sample type F (formed base material) showing higher mass loss than sample type A (unformed base material). Similar behavior was observed between sample C-FSW and sample type D-FSW + forming with sample type D showing higher mass loss than sample type C. Overall, all the sample types showed mass loss below 15 mg/cm$^2$ indicating resistance toward IGC.

### 4.15.4 Pitting Corrosion Testing Results

Pitting corrosion testing was performed per ASTM G 46 (1994). Pitting corrosion of type A (unformed base material), C (friction stir welded material), and E (gas metal arc welded material) was carried out by immersing the samples in 3.5 wt% sodium chloride solution for duration of 1 week. The pit shape was analyzed as per ASTM G46.

Figure 4.107A and B shows macrographs of pit surface at 50× magnification of a corroded surface. The cross-sectional macrograph at 300× magnification for type A, C, and E are shown in Figure 4.107C–G. Wide and shallow pits were observed on type A

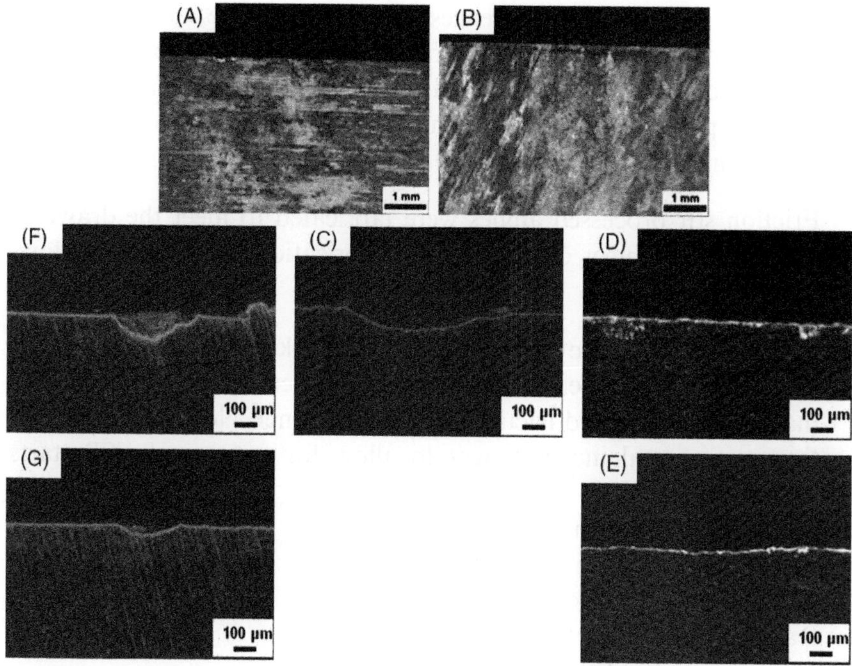


*Figure 4.107 Macrographs of pit surface at 50× magnification of a corroded surface.*

material. Type E material showed more elliptical type of pits as compared to more uniform attack observed in type C material. The depth of attack for type A and E samples was observed to be in similar range (~100–200 μm), whereas type C showed a shallower attack.

## 4.15.5 Crevice Corrosion Testing Results

Crevice corrosion testing was reviewed with reference to ASTM G 48 (2003) and ASTM G78 (2001). These are specifications for testing of steel. With that, there are no specifications for crevice corrosion testing of aluminum. With this, no crevice corrosion testing results for 5083 could be performed.

## 4.15.6 Natural Exposure Testing Results

To perform the natural exposure testing, an actual application was selected for which representative articles could be installed. The US Navy's Littoral Combat Ship program was selected for natural exposure testing. Angles were selected that would be exposed to the natural seawater atmosphere environment on the external portion of the ship. These angles that were fabricated and installed are being periodically inspected to observe any differences (positive, negative, or neutral) versus the traditional GMAW approach. Inspection of both the base material and gas metal arc welded regions is being performed. As of the time of publication, no pertinent data exists due to the limited natural exposure time for the angles of interest.

Friction stir processed angles were fabricated to meet the drawings available for the angles of interest. The fabrication involved the following operations:

1. Procurement of plate of the appropriate thickness
2. Plasma cutting of exterior shape and other features required to mate with the related friction stir welded panel. Note, the plate section was several inches longer to allow for remove of FSP starts and stops.
3. Marking of FSP/bend line
4. Friction stir processing along the bend line
5. Saw cutting of starts and stops
6. Radiographic inspection of FSP
7. Forming along FSP line
8. Dye-penetrant testing along FSP line
9. Delivery to the customer

The fabricated angle has been installed on the ship and is being periodically inspected per the previous discussion.

### 4.15.7 Corrosion Resistance Testing Results

Corrosion resistance testing was performed per ASTM B 117 (2009). Salt fog testing was carried out for 5083-H111 base material, gas metal arc welded material, friction stir processed material and friction stir welded material. Both formed and unformed samples were cut into rectangular blocks with approx. 120 mm × 40 mm × thickness. All the surfaces of the samples were smoothened using 80 grit silicon carbide paper. Three samples each from different alloy conditions were exposed to salt fog atmosphere for 7 days. Initial mass was measured for all the samples before the exposure. After the test, samples were taken out of salt fog chamber, cleaned with distilled water and allowed to dry. Final mass was measured.

Mass loss for all the alloy conditions was observed to be negligible. Figure 4.108A–E shows photographs of samples after exposure to salt

*Figure 4.108 Photographs of samples after exposure to salt fog chamber for the various sample types (A, E, C, F, and D).*

fog chamber for the various sample types (A, E, C, F, and D). Over some of the samples discoloration was observed. Overall, the alloy showed resistance toward salt fog environment in all conditions.

### 4.15.8 Exfoliation Corrosion Testing Results

Exfoliation corrosion testing was performed per ASTM G66 [4.8.20]. Exfoliation is a form of corrosion that proceeds laterally from the sites of initiation along planes parallel to the surface, generally at grain boundaries, forming corrosion products that force metal away from the body of the material, giving rise to a layered appearance. ASTM G66 was followed to estimate the exfoliation corrosion resistance of the various sample types in the 5083-H111. Samples were exposed to

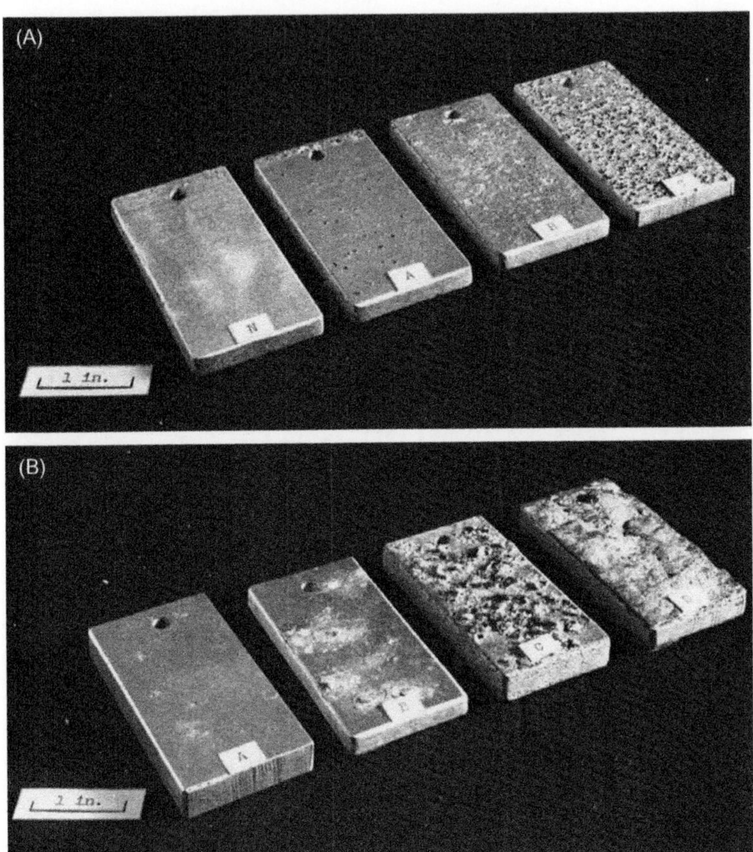


*Figure 4.109 Photographic comparison with the standard chart provided in ASTM G66.*

test solution of 1.0 M ammonium chloride, 0.25 M ammonium nitrate, 0.01 M ammonium tartrate, and 0.09 M hydrogen peroxide at 65°C temperature for a duration of 24 h.

ASTM G66 includes photographic comparison charts to assess the extent and type of corrosion. The photographic comparison with the standard chart provided in ASTM G66 is shown in Figure 4.109A and B and the actual results are shown in Figure 4.110. Pitting type corrosion is shown in the left figures with various extents indicated by letters A–D. Similarly, exfoliation type and extent is shown in the right. Based on the chart comparison it was observed that base material showed pitting type B, GMAW sample showed pitting type A mainly concentrated at the interface between nugget and parent metal,

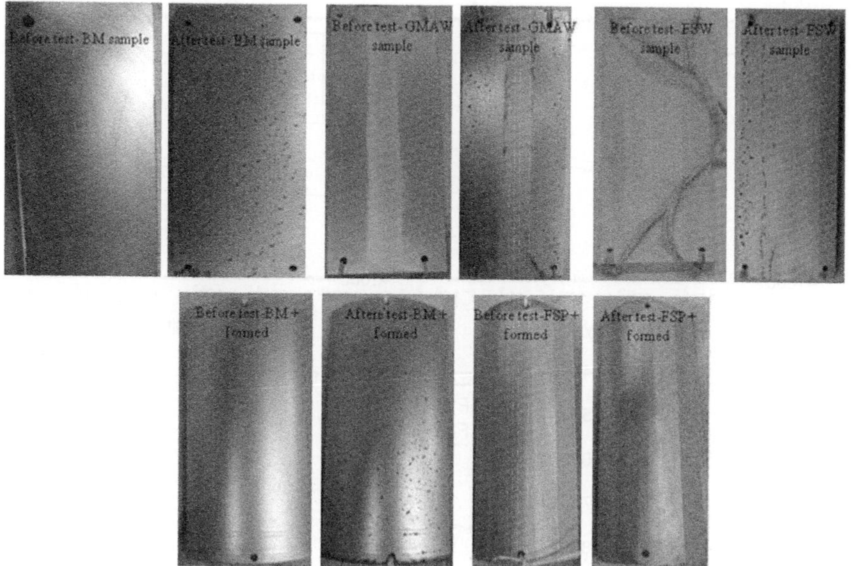


*Figure 4.110 Photographic results of different specimens tested according to ASTM G66.*

whereas FSW stir zone region showed no appreciable pitting attack. However, the retreating side of FSW sample showed pitting type A with pits aligned in a row. Formed base metal samples showed similar pitting behavior as the unformed base metal samples. However, formed base metal samples showed marginally deeper pitting. The formed FSW samples showed pits on the retreating side only. The pits were not in a straight line as observed in FSW unformed samples. The formed FSP region showed better corrosion behavior than formed BM. Overall, none of the sample type showed susceptibility toward exfoliation corrosion with FSW stir zone indicating better corrosion resistance.

### 4.15.9 Electrochemical Corrosion Testing Results

Electrochemical corrosion testing is a relatively rapid technique to estimate the corrosion response of a material when exposed to a particular environment. Material response in terms of potential and current gives indication of corrosion behavior. Linear potentiodynamic polarization technique was implemented. Samples were prepared to estimate the electrochemical parameters of AA 5083 base material with different alloy conditions. Polarization studies were carried out in 3.5 wt% sodium chloride at room temperature. Flat cell with three electrodes

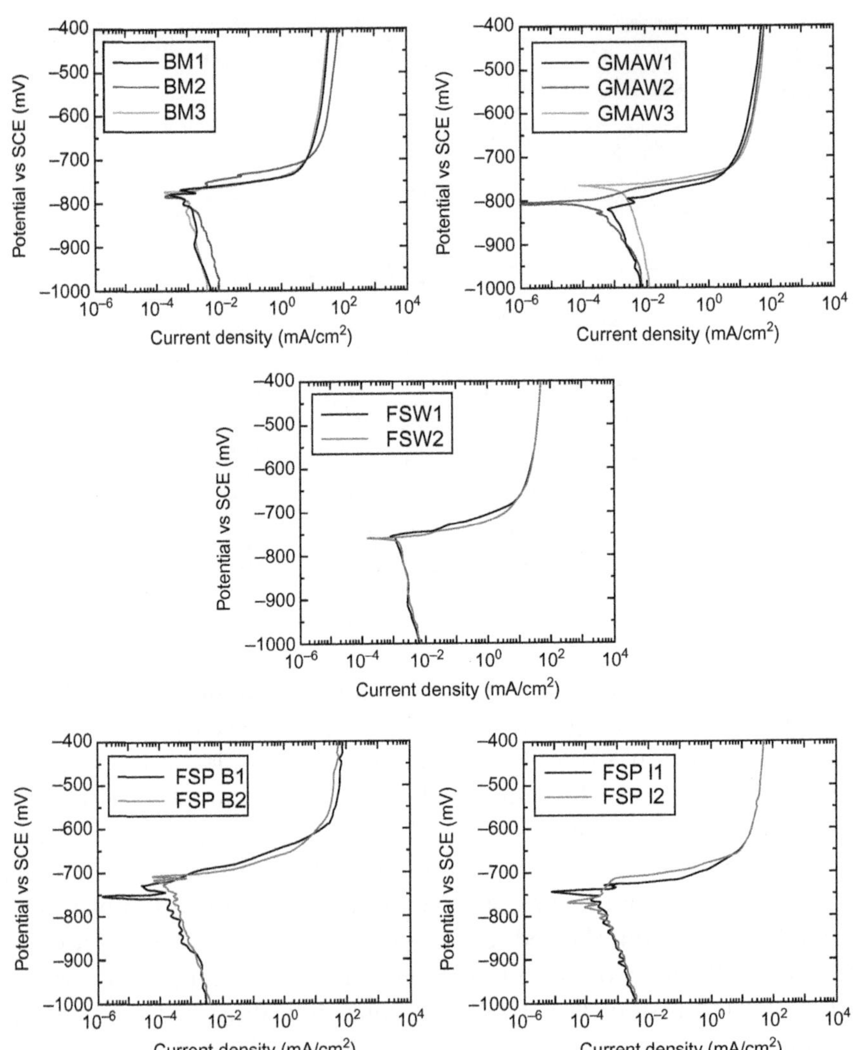


*Figure 4.111 Linear polarization plots of different types of samples.*

configuration was used with sample as working electrode, platinum gage as counter electrode and saturated calomel as reference electrode. Samples were given induction period of 1 h in the test solution before starting the experiment. Samples were scanned from −1000 to −400 mV at a scan rate of 0.166 mV/s, per ASTM-G61.

Figure 4.111 shows linear polarization plots of different types of samples. $E_{corr}$ for FSW sample was found to be positive as compared to base material and GMAW samples. The current density for FSW sample was found to be of one order more than the base metal and GMAW samples. It was observed that FSP samples with both type B and I showed tendency of small passive region in the potential range approximately −750 to −700 mV versus SCE. The current density observed in these samples was 1 to 2 orders less than base metal and GMAW samples. The higher current densities shown by the FSW samples could be attributed to the fact that along with nugget region, small area of TMAZ might have got exposed during test. However, FSP samples with only nugget region showed lower current densities than the base material and GMAW.

## 4.16 FATIGUE TESTING

### 4.16.1 Fatigue Testing Background

Fatigue life is a combination of crack initiation and subsequent crack propagation. For low load high cycle fatigue, fatigue life is dominated by the crack initiation phase, that is, most of the fatigue cycles are associated with crack initiation. Unless there are subsurface defects, fatigue crack initiation usually occurs on the surface and is most often associated with a surface stress concentration such as an edge, weld toe, or other such discontinuity. Thus, surface finish is important. For the test samples fabricating using FSP, the surface in the noted data was ground to eliminate the surface tool marks associated with the rotating FSP tool. In addition, any flash was removed. Thus, following FSP and before bending, the surface is to be smooth with no potential geometric discontinuities. In the absence of surface geometric discontinuities, and with all else equal, crack initiation generally can be directly associated with the metal yield strength. Accordingly, any changes in yield strength can affect fatigue life.

Figure 4.112 shows microtensile properties for the different locations for friction stir processed 5083-H116 plate following FSP. Locations 1 and 5 are parent metal properties. Yield strength for the parent metal varies from approximately 290 to 300 MPa. These are the lowest yield strengths that were noted when compared to samples within and adjacent to the FSP zone. The FSP zone (location 3_1) has a yield strength >360 MPa while the HAZ (locations 3_2 and 3_4) and the TMAZ (locations 2 and 4) have yield strengths >330 MPa. Thus, for 5083-H116, the yield strength in the FSP and adjacent zones

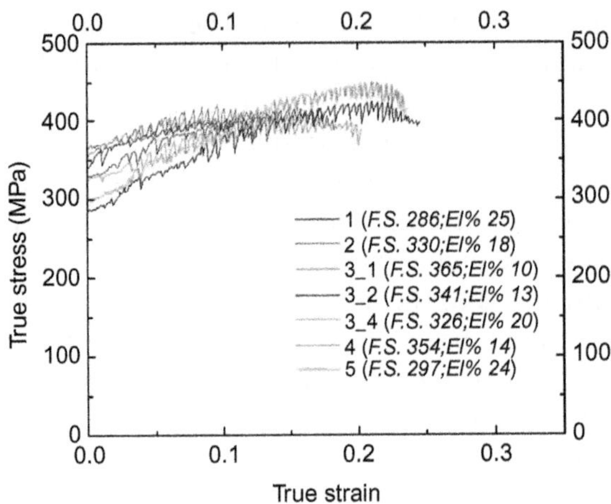


*Figure 4.112 Microtensile properties for the different locations for friction stir processed 5083-H116 plate following FSP.*

is higher than parent metal. This would be expected from the hardness results, that is, higher hardness reported in the FSP zone. Extrapolating yield strength to fatigue, one would anticipate a fatigue life equal to or greater than the parent metal. Unfortunately, data for fatigue resistance in friction stir welded AA 5083 are sparse. However, Tovo et al. (2000) refers to an additional reference wherein it is stated that "friction stir welds in AA 5083 have a high fatigue resistance, comparable to that of the base alloy."

There are additional fatigue life data ($R = -1$) reported for AA 5083 by James et al. (2002), but the temper is H321. Thus, for this temper, a reduced hardness, yield strength, and corresponding fatigue life would be expected compared to parent metal. Unfortunately, the authors did not report parent metal fatigue life. However, the data are informative with regard to surface finish. Fatigue life tests were run for both as-welded (flash removed but surface tool marks not removed) and polished (flash and surface tool marks removed). Referring to the data, if $10^7$ cycles is selected for comparison, the stress amplitude for failure for as-welded samples is $\sim$80 MPa compared to $>$120 MPa for polished samples. This result highlights the benefits of the smooth surface finish. However, it cannot be concluded from this result that the surface must be polished. The weld concavity for the work reported by James et al. (2002) was $\sim$0.2 mm deep. With the newer FSP design the concavity was less in the case study data.

In a separate study, four-point bend fatigue studies were also completed using 5083-H321 comparing fatigue life for a gas metal arc weld and the same weld friction stir processed to a shallow depth on the surface (Fuller et al., 2003). The addition of FSP to the GMAW crown significantly increased the fatigue life. A run-out specimen (no failure after $10^7$ cycles) for the as-GMAW sample was reached at a load of 46 kg, while the FSP sample produced a run-out at 60 kg, a 30% increase in applied load.

FSP increases the hardness and yield strength in 5083-H116. If the FSP surface is polished, fatigue life would be expected to be equal to or better than the parent metal. Equivalent fatigue life for friction stir welded AA 5083 compared to the base metal has also been reported. Based on the limited data available, and the measured mechanical properties, prior to initiation of the case study it was expected that the fatigue life of the FSP and FSW samples would exceed that of the traditional GMAW approach.

### 4.16.2 Mini Fatigue Testing Results

Fatigue testing was performed per ASTM E 466 (2007) and reported per ASTM E 468 (1990). In this work, fatigue tests were performed on the base material, FSP material with two different tool designs, GMAW material, formed base material, formed FSP material, FSW material by using mini fatigue samples (as shown in Figure 4.113) under completely reversed flexure ($R$ ratio = −1).

The bending stress in the specimens was uniform across the test section and completely reversed ($R$ ratio = −1). Three stress ratio levels

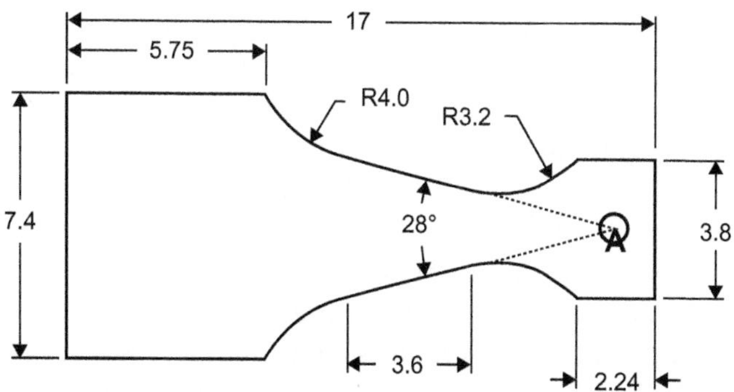


*Figure 4.113 Mini fatigue specimens used in this study.*

**Table 4.3 Summary of Yield Stress, Applied Stress, Amplitude Values, and Corresponding Fatigue Life of Each Material**

| Sample | YS (MPa) | Stress Amp. (MPa) | $N_f$ |
|---|---|---|---|
| A (base material) | 229 | 215 | $1.41 \times 10^5$ |
| | | 197 | $2.48 \times 10^5$ |
| | | 187 | $1.65 \times 10^6$ |
| B (FSP—scroll shoulder tool) | 166 | 157 | $6.97 \times 10^5$ |
| | | 148 | $1.70 \times 10^6$ |
| | | 141 | $2.52 \times 10^6$ |
| E (GMAW) | 132 | 125 | $7.12 \times 10^5$ |
| | | 120 | $1.70 \times 10^6$ |
| | | 110 | $2.75 \times 10^6$ |
| I (FSP—stepped spiral tool) | 185 | 176 | $9.80 \times 10^5$ |
| | | 167 | $2.37 \times 10^6$ |
| | | 157 | $4.55 \times 10^6$ |
| F (base material + forming) | 263 | 252 | $7.90 \times 10^4$ |
| | | 237 | $1.40 \times 10^5$ |
| | | 222 | $9.07 \times 10^5$ |
| D (FSP + forming—stepped spiral tool) | 273 | 261 | $1.50 \times 10^5$ |
| | | 244 | $5.90 \times 10^5$ |
| | | 230 | $1.01 \times 10^6$ |
| C (FSW) | 163 | 155 | $8.6 \times 10^5$ |
| | | 147 | $2.70 \times 10^6$ |
| | | 128 | $8.30 \times 10^6$ |

of ~0.95, 0.9, and 0.85 of yield stress (YS) were designed for the constant amplitude fatigue tests for each material. It is noted that the yield stress values were from mini tensile tests for top layer of each material. The yield stress, applied stress amplitude values and corresponding fatigue life of each material are given in Table 4.3. Figure 4.114 shows the fatigue results represented by S–N curves. For comparison, fatigue data from handbook for 5083Al-O and 5083Al-H116 were also referred in this figure. It can be seen that among the unformed samples, base material exhibits highest fatigue strength. However, at the same percentage of YS level, a significant increase in fatigue life obtained in FSP, FSW, and GMAW samples is observed. Compared to GMAW sample, FSP samples have a similar fatigue life at the same

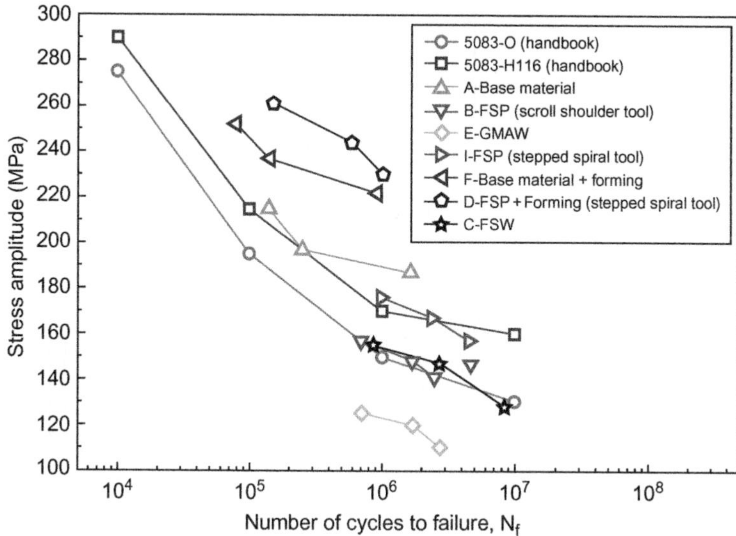


*Figure 4.114 The fatigue results represented by S–N curves for various materials.*

percentage of YS level, but show higher fatigue strength. For instance, corresponding to a fatigue life of $10^6$ cycles, the fatigue strength for FSP samples (B and I) and GMAW sample are ~152, 180, and 123 MPa, respectively. Comparing FSW with GMAW, both higher fatigue strength and better fatigue life were obtained in the FSW samples. It is also noted that the FSP samples processed by different tools show different fatigue properties. The higher fatigue properties are likely correlated with a finer or/and more uniform microstructure in the processed material. The FSP and FSW samples show comparable fatigue properties with the materials 5083Al-O and 5083Al-H116.

For the formed materials, they have higher fatigue strength than the unformed materials. Compared with the formed base material, the formed FSP material shows higher fatigue strength and even a better fatigue life. The present results indicate that a good combination of fatigue strength and fatigue life can be obtained in the FSP samples. This may be attributed to the grain refinement of the matrix and the breakup of coarse constituent particles during FSP.

### 4.16.3 Standard Fatigue Testing

In this work, standard force controlled constant amplitude axial fatigue tests were performed for the base material, FSW and GMAW

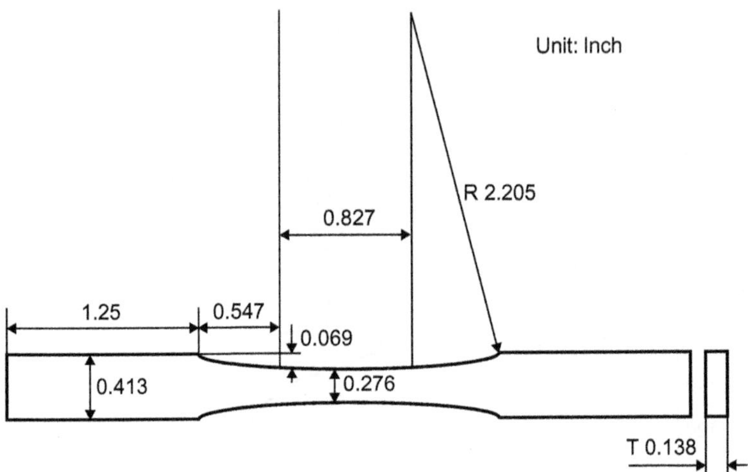


*Figure 4.115 Large fatigue specimen geometry used for this study.*

using ASTM E466-07 fatigue specimens, as shown in Figure 4.115. Due to the sample size, only coupons with flat geometry (unformed) could be tested. Since, the sample requires homogenous material, fatigue testing of the FSP sample was not possible. However, since FSW requires more heat input, FSP results would only be expected to be better than FSW.

### 4.16.4 Results of Standard Fatigue Testing Results

The fatigue tests were performed under tensile–tensile loading with $R = 0.1$. Three stress ratio levels of ~0.95, 0.9, and 0.85 of yield stress (YS) were tested for the constant amplitude axial fatigue tests for each material. The yield stress values are the average value of top three layer mini tensile samples of each material. The yield stress, applied stress amplitude values and corresponding fatigue life of each material are given in Table 4.4. Figure 4.116 shows the fatigue results represented by S–N curves.

In general, the standard ASTM E466-07 axial fatigue tests show similar tendency with the mini bending fatigue tests. Among the testing materials, the base material shows highest fatigue strength and the GMAW sample has the lowest fatigue strength. Compared with GMAW sample, FSW sample shows higher fatigue strength and better fatigue life. It is noted that for GMAW sample, a relatively low fatigue life was obtained at 0.9 YS (123 MPa) level. In that case, one more

**Table 4.4 Summary of Yield Stress, Applied Stress, Amplitude Values, and Corresponding Fatigue Life of Each Material**

| Sample | YS (MPa) | Stress Amplitude (MPa) | $N_f$ |
|---|---|---|---|
| A (base material) | 243 | 231 | $2.29 \times 10^5$ |
| | | 219 | $9.08 \times 10^5$ |
| | | 207 | $1.23 \times 10^6$ |
| C (FSW) | 166 | 158 | $2.92 \times 10^6$ |
| | | 149 | $3.70 \times 10^6$ |
| | | 141 | $9.50 \times 10^6$ |
| E (GMAW) | 137 | 130 | $1.83 \times 10^5$ |
| | | 123 | $1.97 \times 10^5$ |
| | | 123 | $5.82 \times 10^5$ |
| | | 116 | $1.83 \times 10^6$ |

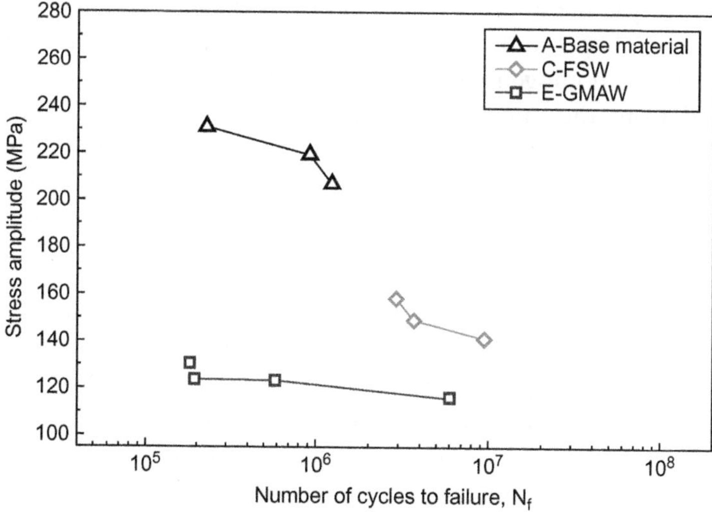


*Figure 4.116 Fatigue data for the large specimens in various conditions.*

sample was tested under same stress level, as shown in Figure 4.116. Analysis of the sample with unexpectedly low life indicated an area of porosity, indicating this data point should be ignored or indicates that fatigue life of GMAW can be expected to be much more variable than that of FSW or FSP which are much less susceptible to porosity type indications (or voids in the case of FSW or FSP)

## 4.17 CORROSION FATIGUE TESTING

### 4.17.1 Background

There were no data in the literature reporting corrosion fatigue life for friction stir welded or welded AA 5083 in any temper. However, based on the evidence for corrosion resistance being equivalent to or better than the parent metal and based on the evidence for fatigue life being equivalent to parent metal, there was no rationale to believe that corrosion fatigue life would be any less for friction stir welded 5083-H116 than the parent metal. Thus, it would be expected that there should be no dramatic differences between the base material and any of the processing techniques.

### 4.17.2 Results

Corrosion fatigue testing was performed per ASTM E 466 (2007) and ASTM F 1801 (1997). Corrosion fatigue in 3.5 wt% sodium chloride solution was conducted for unformed base material, friction stir welded and gas metal arc welded. Only these samples could be tested due to the shape and size requirements of the sample geometry and need for homogenous material. Bending fatigue samples were extracted from the nugget or stir zone region for the gas metal arc welded and friction stir welded samples. Figure 4.117 below shows the picture of corrosion fatigue samples before testing. Corrosion fatigue testing was carried out with an R ratio of −1.0 in the presence of 3.5 wt% sodium chloride solution.

S–N curves for the base material, friction stir welded and gas metal arc welded samples tested at stress level of 0.85, 0.90, and 0.95 fraction of respective yield strength are shown in Figure 4.118. For comparison, base material samples at identical stress levels were also tested in air. For identical levels of stress amplitude/yield stress ratio values at which the corrosion fatigue were carried out, the fatigue life for friction stir welded samples were better or equivalent to type base material samples. Figure 4.119 shows macro images of fracture surfaces of failed samples. For type base material and friction stir welded samples tested in solution fine cracks were observed in the region away from failed surface indicating influence of corrosive medium on the surface.

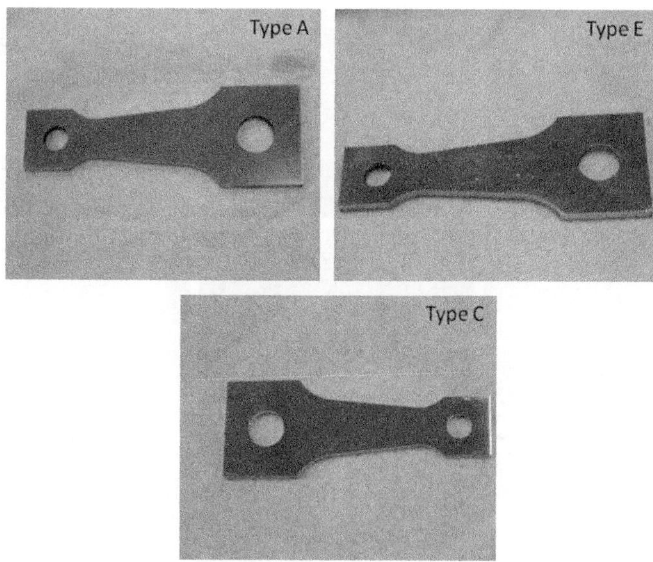


*Figure 4.117 Photographs of corrosion fatigue samples before testing.*

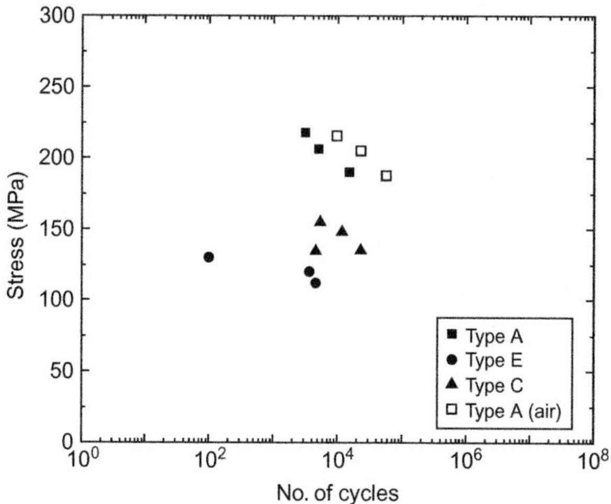


*Figure 4.118 S–N curves for the base material, friction stir welded and gas metal arc welded samples tested at stress level of 0.85, 0.90, and 0.95 fraction of respective yield strength.*

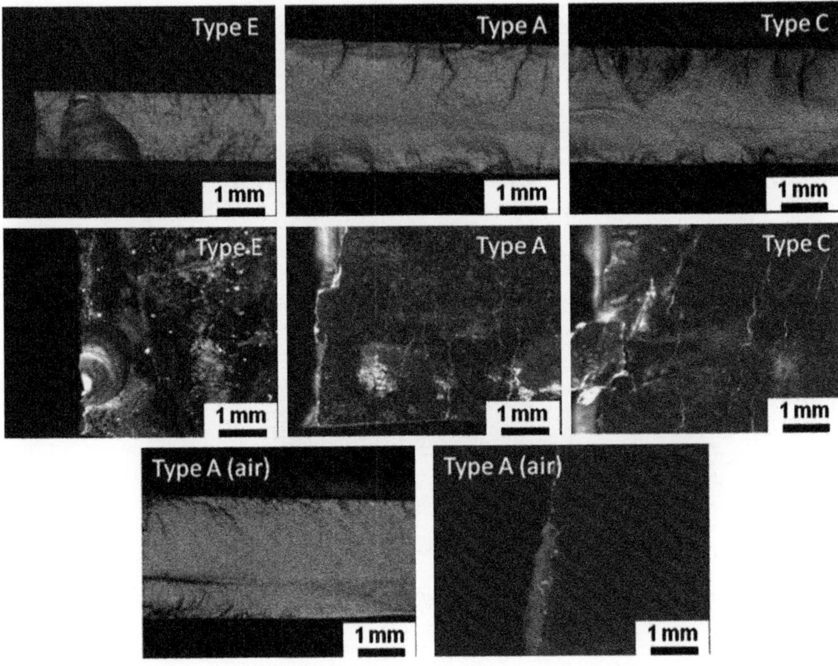


*Figure 4.119 Macro images of fracture surfaces of failed samples.*

For gas metal arc welded samples, the porosity in the gage region led to early failure of samples and enough duration was not available for corrosive medium to influence the surface. The porosity in the gas metal arc welded samples also led to wide scatter in corrosion fatigue life. For base material samples tested in absence of solution (in air), no fine crack features in the region away from failed surface were observed.

CHAPTER 5

# Examples of Enhanced Formability of High-Strength Aluminum Alloys

## 5.1 BACKGROUND

The previous chapters have summarized data from a large case study of application of friction stir processing (FSP) to 5083 alloy to locally improve its ductility and subsequently provide improved material properties versus an equivalent fabrication joined via traditional fusion welding processes. All of the FSP that was performed in the subject of the case study at this point was a single-friction stir pass. With this, there are a significant number of other application variants where FSP can be used to locally improve the ductility of aluminum alloys. These other applications can generally be summarized as being on thicker material where typically high-strength aluminum alloys are used in structural applications. In these other applications, the area of aluminum that needs to be locally modified tends to be larger due to the forming areas necessarily being larger and greater minimum bend radii. Given this situation, many of the additional potential applications require FSP with multiple passes, so as to cover a larger area. These additional application areas are the subject of this chapter.

## 5.2 EXAMPLES OF ENHANCED FORMABILITY OF HIGH-STRENGTH ALUMINUM ALLOYS

The first application that is reviewed is FSP of thick section 6061-T6. 6061-T6 is a structural alloy that is heat treatable and has relatively low ductility. There are very few applications where it is formed, due to the low ductility and the fact that it is relatively easy to extrude. Thus, complex shapes are often produced by the extrusion process versus considering any sort of forming process.

In thick section 6061-T6, FSP to enable or enhance forming has been performed on material ranging from 25 to 150 mm (Mahoney et al., 2005). In one example in the cited reference, FSP of a 25 mm thick plate was performed to a depth of 6 mm. Subsequently, the plate

Friction Stir Processing for Enhanced Low Temperature Formability. DOI: http://dx.doi.org/10.1016/B978-0-12-420113-2.00005-2

was machined down 3 mm to remove any surface tool marks, which could have acted as stress risers. To compare the effectiveness of the FSP to enhance forming, plates with and without FSP were bent. The authors reported that the unprocessed plates could only achieve a bend radius of approximately 230 mm (9 in.) before failing. The plates failed at about a 25° bend angle. Comparatively, the friction stir processed sample was able to attain a bend angle in excess of 80°. Figure 5.1 displays the processed and unprocessed samples after forming. Because the 6061-T6 alloy is in a heat-treated condition, the locally improved ductility was primarily attributed to the annealing effect of the FSP due to the process' inherent heat input. Tension tests of material near the surface show the dramatic improvement in ductility. This is shown in Figure 5.2. The figure also shows that the ultimate tensile strength and yield strength have been reduced. However, equal to or near parent material properties could be regained by a post forming heat treatment process.

The same authors also tested the effects of FSP on much thicker 6061-T6. The testing included FSP of 150 mm thick material to a depth of 25 mm. In this particular investigation, a multi-pass FSP pattern (referred to as a raster pattern) in a spiral path or shape was performed. As in the previous work, a small amount of material (6 mm) was machined off the surface to remove any potential surface effects. Similarly, post FSP bending of the samples was performed. The friction stir processed sample was able to attain a bend angle of 30° (maximum extent of bending die), whereas an unprocessed sample was only able to attain a bend angle of 8° before failing. The friction stir processed and formed 150 mm thick sample is shown in Figure 5.3.

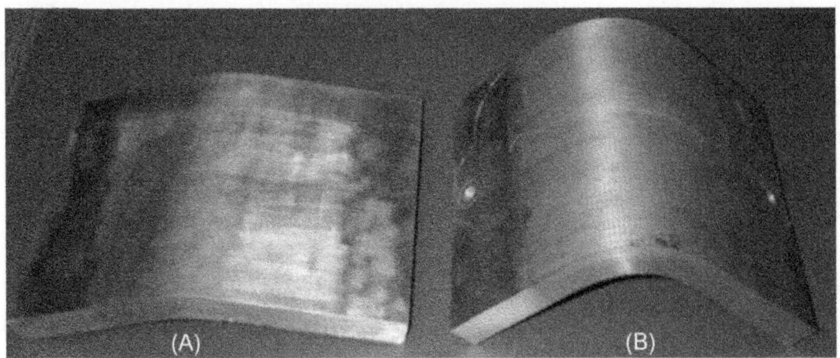


*Figure 5.1 Plane strain bending in 50 mm thick 6061-T6 Al. (A) Parent metal bent to 27° with cracks initiating on the tensile surface and (B) FSP 6061-T6 Al bent to 85° without cracking* (from Mishra and Mahoney, 2007).

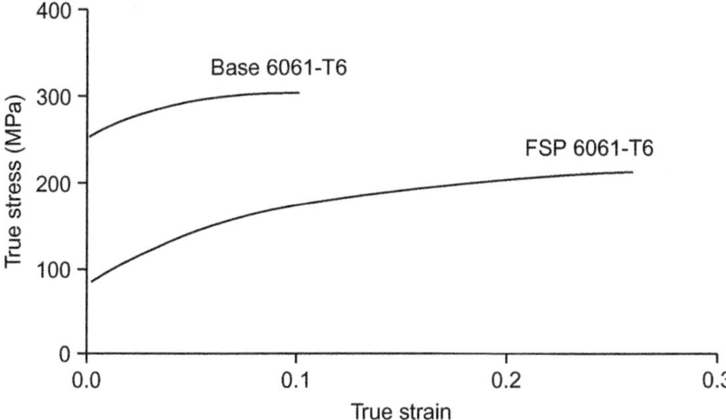


*Figure 5.2 True stress–true strain tensile curves for base 6061-T6 and friction stir processed 6061-T6 (5 mm depth of processing). The area under the base material curve is 29 MPa, while the area under the FSP material curve is 45 MPa* (from Mishra and Mahoney, 2007).

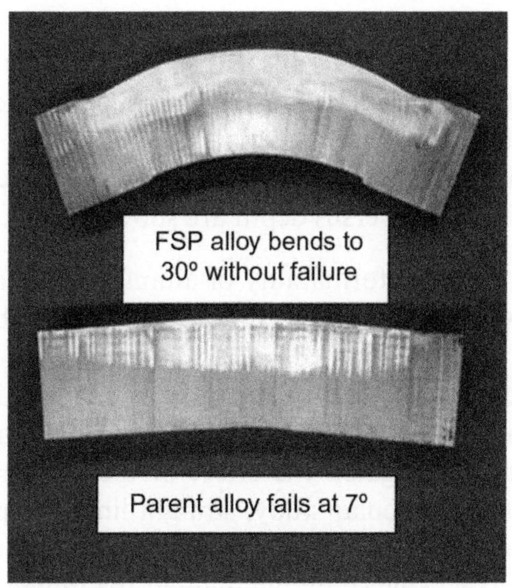


*Figure 5.3 Friction stir processed 150 mm (6 in.) thick 6061-T6 Al bent to 30° without cracking compared to parent metal reaching a bend limit of 7°* (from Mishra and Mahoney, 2007).

Similar FSP trials have been performed on other high-strength alloys. One such example is 7050-T7451 (Mahoney et al., 2001, 2002). In this study, 50 mm thick material was investigated, with FSP in a couple of different raster patterns to a depth of 6 mm. The raster patterns included a spiral and a zigzag. The purpose of this different raster pattern was to investigate the difference between overlapping common sides of the previous pass (spiral) versus overlapping opposite sides (zigzag). In the former, each pass overlapped the previous pass' advancing side, resulting in a more homogenous microstructure throughout the processed area. Similarly, forming trials were performed after FSP. With this particular alloy and the selected processing parameters, it was demonstrated that the spiral raster pattern yielded a more homogenous microstructure, and enabled bending to a greater diameter before failure. An example of a formed sample is shown in Figure 5.4 and a cross-section is shown in Figure 5.5.

As was also performed with other alloys, tensile and hardness tests were performed on the thick plate 7050-T7451 alloy friction stir processed samples. Testing was performed at various depths from the surface. Given the fact that FSP necessarily inputs heat into the material, the hardness, ultimate tensile strength, and yield strength are negatively impacted by the FSP. However, since the FSP was only performed to a depth of 6 mm, all of these properties increased as the test location depth increased. Contrarily, the ductility increases near the surface compared to the base material but decreases toward the base material properties versus the depth after FSP. The processed materials stress versus strain curves versus depth are shown in Figure 5.6.

The ability to enhance formability of aluminum has also been investigated on 2XXX aluminum alloy. In one such study, the effect of friction stir welding was investigated on 25 mm thick 2519-T87 (Mahoney et al., 2001). The authors took a similar approach in this investigation, as well. In this particular work, the FSP depth was 6 mm. As with the 7050, the authors investigated the effect of the inhomogeneity of the FSP process. In this particular study, straight line FSP passes were performed where the direction of the FSP was either parallel or perpendicular to the bend axis. This study highlighted the inhomogeneity, as the samples where the FSP was perpendicular to the bend axis performed better in the forming trials, with greater bend angles before failure being observed. Figure 5.7 shows a sample that was formed to a bend angle of 85° and clearly shows the extent or depth to which the FSP was performed.

As with the other studies, microhardness data was measured as a function of depth. As would be expected and as is shown in Figure 5.8,

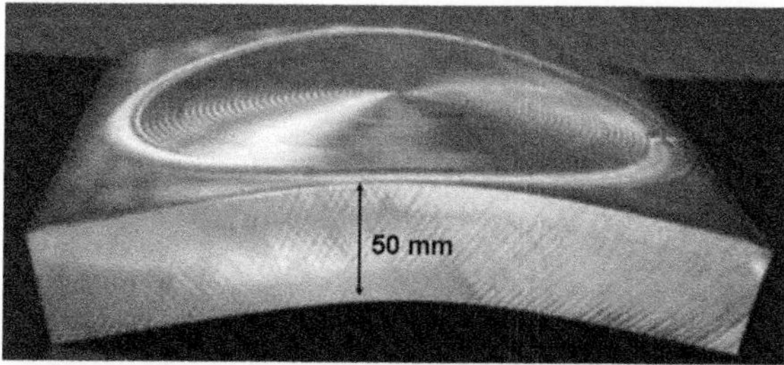


*Figure 5.4 Spiral raster pattern in 50 mm thick FSP 7050-T7451 Al bent 16° at room temperature* (from Mishra and Mahoney, 2007).

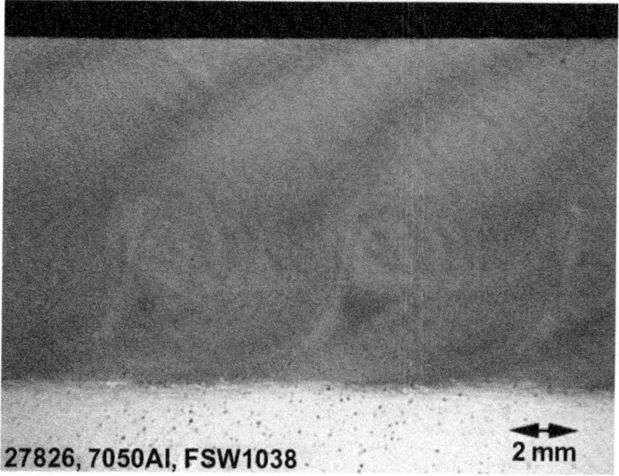


*Figure 5.5 Longitudinal cross-section illustrating depth and deformation pattern in the FSP zone* (from Mishra and Mahoney, 2007).

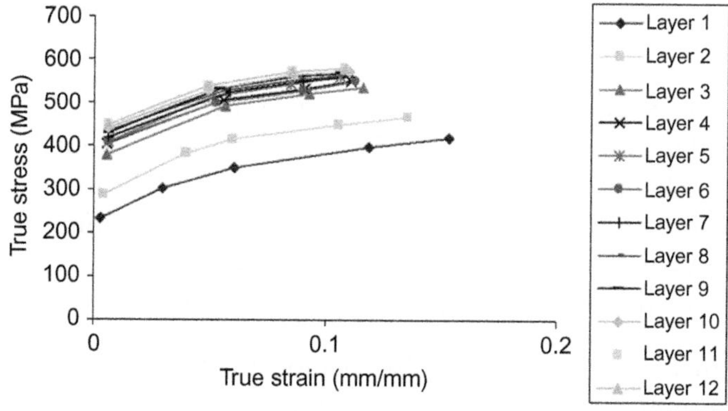


*Figure 5.6 True stress–true strain tensile curves for layers taken through the thickness of a 50 mm thick 7075 friction stir processed plate. Layer 1 consists entirely of friction stir processed material, while layer 12 is on the opposite side of the plate. The area under the curve for layer 1 is 58 MPa and the areas under the other curves decrease more or less uniformly to about 54 MPa in layer 12* (from Mishra and Mahoney, 2007).

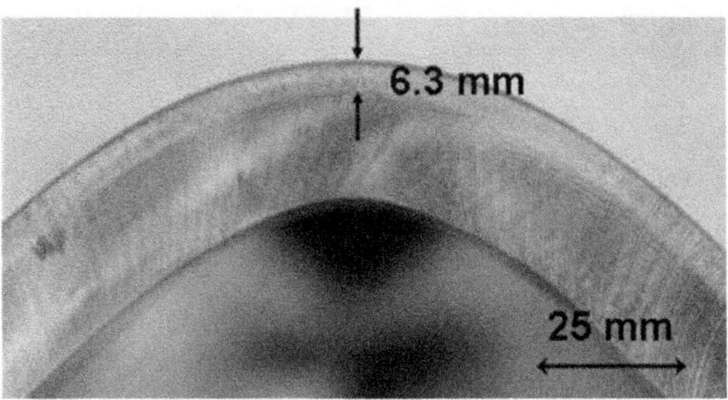


*Figure 5.7 Illustration of the FSP depth (6.3 mm) and the ability to bend 2519-T87 Al ~85° at room temperature* (from Mishra and Mahoney, 2007).

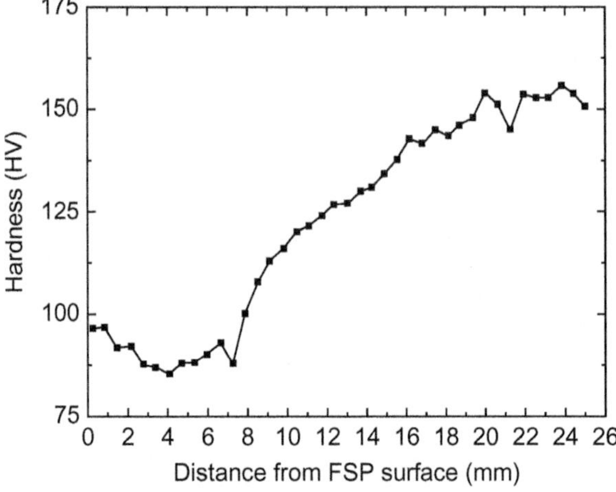


*Figure 5.8 Hardness in friction stir processed 2519 as a function of depth below the surface. Note the relatively deep heat-affected zone (15–18 mm)* (from Mishra and Mahoney, 2007).

the hardness does increase with depth, given the fact that this is also a heat treatable alloy and that the FSP depth was only about 25% of the material thickness. This is similar to the other alloys. Additionally, tensile tests were performed to gain some insight into the improvement in ductility. As with the other alloys, the stress–strain curve shown in Figure 5.9 readily highlights the improvement in the ductility.

A further enhancement is also demonstrated with this particular alloy. The enhancement included machining after the bending operation. This is shown in Figure 5.10. As can be seen, the outer surfaces have been machined to a corner, with reduction of thickness in the remaining areas. Machining to shapes as shown in Figure 5.10 can

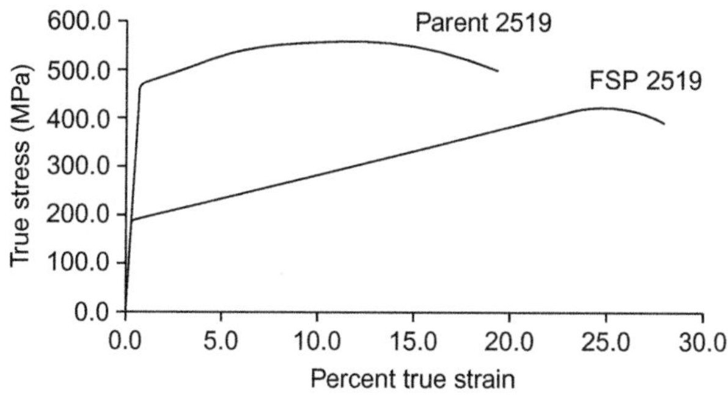


*Figure 5.9 True stress versus true strain for as-received 2519-T87 Al and following FSP* (from Mishra and Mahoney, 2007).

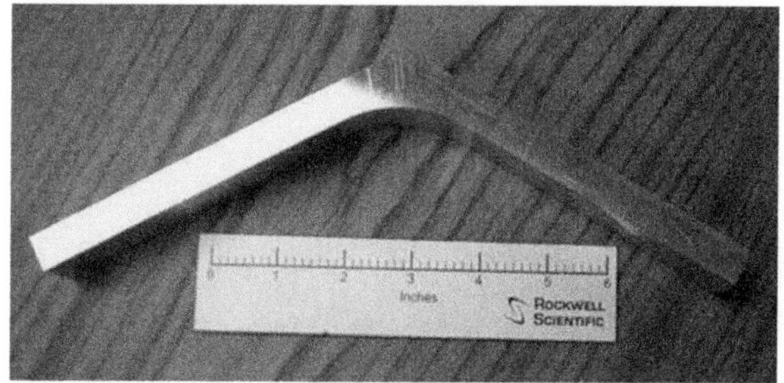


*Figure 5.10 Final structure following room temperature bending and machining to final thickness following FSP* (from Mishra and Mahoney, 2007).

have benefits in stealth applications, where parts with sharp corners tend to have improved stealth capabilities or reduced radar signatures. This is obviously a potential significant benefit in military applications. Another benefit is that this particular approach could avoid the use of traditional welding to fabricate structures where plate sections are joined at various angles to one another. With this, the improved properties of FSP would yield improved ballistic properties as well as the other properties demonstrated in the previous chapters.

## 5.3 SUMMARY

As was demonstrated throughout the previous chapters, FSP can provide significant benefits in the formability of thin plate aluminum. However, it is not just thin plate where FSP can have significant benefits. FSP has also been demonstrated to enhance the ductility of thick section aluminum, as described in this chapter. Such applications will typically require multiple friction stir process passes to process enough material to cover the entire bend area. This addition of multiple passes was shown in multiple cases to highlight the inhomogeneity of the FSP processes and highlights the fact that the direction, shape, and orientation of the FSP in situations where multiple passes are required, which is an important variable that must be considered. Limited data so far suggests that FSP can be used as a manufacturing technology to enhance formability and provide superior post process material properties in thick section high-strength aluminum alloys versus traditional joining approaches. In fact, a significant percentage of these high-strength aluminum alloys are unweldable with traditional fusion welding technologies. Thus, FSP could be an enabling technology for fabrications where thick section high-strength aluminum alloys would be of benefit but heretofore could not be considered.

# REFERENCES

## CHAPTER 1

Dawes, C., Thomas, W., 1995. Friction Stir Joining of Aluminium Alloys. TWI Bull. 6, 124.

Mishra, R.S., Ma, Z.Y., 2005. Friction Stir Welding and Processing. Mater. Sci. Eng. R 50, 1.

Mishra, R.S., Mahoney, M.W., 2001. Friction Stir Processing: A New Grain Refinement Technique to Achieve High Strain Rate Superplasticity in Commercial Alloys. Materials Science Forum, vol. 357–359. Trans Tech Publications, Switzerland, pp. 507–514.

Mishra, R.S., Mahoney, M.W., 2007. Friction Stir Processing, in Friction Stir Welding and Processing. ASM International (Chapter 14) pp. 309–350, Materials Park, Ohio.

Mishra, R.S., Mahoney, M.W., McFadden, S.X., Mara, N.A., Mukherjee, A.K., 1999. High Strain Rate Superplasticity in a Friction Stir Processed 7075 Al Alloy. Scr. Mater. 42, 163–168.

Thomas, W.M., Nicholas, E.D., Needham, J.C., Murch, M.G., Templesmith, P., Dawes, C.J., G. B. Patent Application No. 9125978.8, December 1991.

## CHAPTER 2

Datsko, J., Yang, C.T., 1960. Correlation of Bendability of Materials With Their Tensile Properties. Trans. ASME B82 (4), 309–313.

Ghosh, A.K., Hecker, S.S., 1974. Stretching Limits in Sheet Metals: In-plane versus out-of-plane Deformation. Metall. Trans. 5, 2161–2164.

Kalpakjian, S., Schmid, S.R., 2006. Manufacturing Engineering and Technology, Fifth Ed. Pearson Education, Inc., Upper Saddle River, NJ, ISBN 0-13-148965-8.

## CHAPTER 3

Kalpakjian, S., Schmid, S.R., 2006. Manufacturing Engineering and Technology, Fifth Ed. Pearson Education, Inc, Upper Saddle River, NJ, ISBN 0-13-148965-8.

## CHAPTER 4

Aluminum Standards and Data 2009. The Aluminum Association, 2009.

ASTM B 117–2009 Standard Practice for Operating Salt Spray (Fog) Apparatus, American Society of Testing of Materials, July 2009.

ASTM G 66–1999 (E 2006) (R 2005) Standard Test Method for Visual Assessment of Exfoliation Corrosion Susceptibility of 5XXX Series Aluminum Alloys (ASSET Test), American Society of Testing of Materials, April 1999.

ASTM E 340–2000 (R 2006) Standard Test Method for Macroetching Metals and Alloys, American Society for Testing of Materials, May 2000.

ASTM E 340–2009 Standard Test Method for Micro-Indentation Hardness of Materials, American Society of Testing of Materials, May 2009.

ASTM E 466–2007 Standard Practice for Conducting Force Controlled Constant Amplitude Axial Fatigue Tests of Metallic Materials, American Society of Testing of Materials, November 2007.

ASTM E 468–1990 (E 2004) (R 2004) Standard Practice for Presentation of Constant Amplitude Fatigue Test Results for Metallic Materials, American Society of Testing of Materials, April 1990.

ASTM E 756–2005 Standard Method for Measuring Vibration-Damping Properties of Materials, American Society of Testing of Materials, 2005.

ASTM F 1801–1997 (R 2004) Standard Practice for Corrosion Fatigue Testing of Metallic Implant Materials, American Society of Testing of Materials, April 1997.

ASTM G 46–1994 (R 2005) Standard Guide for Examination and Evaluation of Pitting Corrosion, American Society of Testing of Materials, February 1994.

ASTM G 48–2003 (R 2009) Standard Test Methods for Pitting and Crevice Corrosion Resistance of Stainless Steels and Related Alloys by Use of Ferric Chloride Solution, American Society of Testing of Materials, May 2003.

ASTM G 49–1985 (R 2005) Standard Practice for Preparation and Use of Direct Tension Stress-Corrosion Test Specimens, American Society of Testing of Materials, June 1985.

ASTM G 67–2004 Standard Test Method for Determining the Susceptibility to Intergranular Corrosion of 5XXX Series Aluminum Alloys by Mass Loss After Exposure to Nitric Acid (NAMLT Test), American Society of Testing of Materials, May 2004.

ASTM G 78–2001 (R 2007) Standard Guide for Crevice Corrosion Testing of Iron-Base and Nickel-Base Stainless Alloys in Seawater and Other Chloride-Containing Aqueous Environments, American Society of Testing of Materials, May 2001.

AWS B4.0: 2007 Standard Methods for Mechanical Testing of Welds, seventh ed. American Welding Society, May 2007.

AWS D1.2: 2008 Structural Welding Code—Aluminum, fifth ed. American Welding Society, June 2008.

Fuller, C., Mahoney, M., Bingel, W., 2003. Friction stir processing of aluminum fusion welds. Proceedings of the Fourth International Symposium on Friction StirWelding, 14–16 May. TWI, Park City, UT.

James, M., Hattingh, D., Bradley, G., 2002. Influence of travel speed on fatigue life of friction stir welds in 5083 aluminum, fatigue. International Fatigue Congress, pp. 413–420.

Kumagai, M., Tanaka, S., Properties of aluminum wide panels by friction stir welding. In: First International Symposium on Friction Stir Welding, June 1999, TWI CD, Thousand Oaks, CA.

Mishra, R.S., 2008. MS&T Mini Tensile Testing Procedure based on ASTM-E8.

NAVSEA #T9074-AX-GIB-010/100, NAVSEA Technical Publication, Material Selection Requirements, January 1999.

NAVSEA #802-7651533A, Friction Stir Welding Fabrication, Welding Workmanship, and NDE Requirements (Project Peculiar Document), November 2006a.

NAVSEA #802-7651532A, Specification for Procedure and Performance Qualification Requirement for Friction Stir Welding of Aluminum (Project Peculiar Document), November 2006b.

Pao, P.S., Fonda, R.W., Jones, H.N., Feng, C.R., Connolly, B.J., Davenport, A.J., 2005. Microstructure, fatigue crack growth, and corrosion in friction stir welded Al 5456. Friction Stir Welding and Processing III. TMS, Warrendale, PA, pp. 27–34.

Prime, M.B., 2001. Cross-sectional mapping of residual stresses by measuring the surface contour after a cut. J. Eng. Mater. Technol. 123, 162–168.

Reynolds, A.P., Mechanical and corrosion performance of TGA and friction stir welded aluminum for tailored welded blanks: alloys 5454 and 6061. In: Proceedings of the Fifth International Conference, Callaway Gardens Resort, Pine Mountain, GA, June 1–5 1998.

Sato, Y., Kokawa, H., 2001. Distribution of tensile property and microstructure in friction stir weld of 6063 aluminum. Metall. Mater. Trans. 32, 3023–3031.

Searles, J.L., Gouma, P.I., Buchheit, R.G., 2001. Stress corrosion cracking of sensitized AA5083 (Al-4.5Mg-1.0Mn). Metall. Mater. Trans. A 32A, 2859–2867.

Second EuroStir Workshop, November 6 2002 TWI, Cambridge, UK.

Smith, C., Mahoney, M., Mishra, R.S., with input from NAVSEA, September 2009. Material System Information (MSI): Friction Stir Processing for Fabrication of Aluminum Structural Angles for Structural Angles and Girders for LCS.

Stuart Bond, Corrosion of Welded Components in Marine Environments, London, April 2–3 2003, Lloyds List Events.

Tovo, R., De Sisciolo, R., Volpone, M., Atti XXIX Convegno Nazionale dell' Associazione Italiana per l'Analisi delle Sol-lecitazioni, Lucca, Settembre 6–9 2000, pp. 39–46.

Vilaca, P., Pepe, N., Quintino, L., Metallurgical and Corrosion Features of Friction Stir Welding of AA5083-H111, Technical University of Lisbon, submitted for publication.

Vitek, J.M., David, S.A., Johnson, J.A., Smartt, H.B., DebRoy, T., Characteristics of friction stir welded aluminum alloys. In: Proceedings of the Fifth International Conference Pine Mountain. GA, June 1–5 1998, pp. 574–579.

Williams, S.W., Price, D.A., Wescott, A., Steuwer, A., Peel, M., Altenkirch, J., et al., 2006. Distortion control in welding by mechanical tensioning. In: Proceeding of the Sixth International FSW Symposium, 10–13 October. TWI, Saint Sauveur, Canada.

Zucchi, F., Trabanelli, G., Grassi, V., 2001. Pitting and stress corrosion cracking resistance of friction stir welded AA 5083. Mater. Corros. 52, 853–857.

## CHAPTER 5

Mahoney, M., Mishra, R.S., Nelson, T., Flintoff, J., Islamgaliev, R., Hovanski, Y., 2001. High Strain Rate, Thick Section Superplasticity Created via Friction Stir Processing, Friction Stir Welding and Processing, November 2001. TMS, Indianapolis, IN.

Mahoney, M.W., Mishra, R., Nelson, T., 2002. High strain rate superplasticity in thick section 7050 aluminum created by friction stir processing. In: Proceedings of the Third International Symposium on Friction Stir Welding, September 2001. TWI, Kobe, Japan.

Mishra, R.S., Mahoney, M.W., 2007. Friction Stir Processing, in Friction Stir Welding and Processing. ASM International (Chapter 14) pp. 309–350.

Mahoney, M., Barnes, A.J., Bingel, W.H., Fuller, C., 2004. Superplastic forming of 7475 aluminum sheet after friction stir processing. Mater. Sci. Forum 447–448, 505–512.

Mahoney, M., Fuller, C., Miles, M., Bingel, W., 2005. In: Jata, K., Mahoney, M., Mishra, R., Lienert, T. (Eds.), Thick Plate Bending of Friction Stir Processed Aluminum Alloys, Friction Stir Welding and Processing III. TMS, pp. 131–137.